MEASUREMENTS OF PHOSPHOR PROPERTIES

MEASUREMENTS OF PHOSPHOR PROPERTIES

Edited by
William M. Yen
Shigeo Shionoya (Deceased)
Hajime Yamamoto

Boca Raton   London   New York

CRC Press is an imprint of the
Taylor & Francis Group, an informa business

This material was previously published in *Phosphor Handbook, Second Edition* © 2007 by Taylor and Francis Group, LLC.

CRC Press
Taylor & Francis Group
6000 Broken Sound Parkway NW, Suite 300
Boca Raton, FL 33487-2742

© 2007 by Taylor & Francis Group, LLC
CRC Press is an imprint of Taylor & Francis Group, an Informa business

No claim to original U.S. Government works
Printed in the United States of America on acid-free paper
10 9 8 7 6 5 4 3 2 1

International Standard Book Number-10: 1-4200-4365-X (Hardcover)
International Standard Book Number-13: 978-1-4200-4365-5 (Hardcover)

This book contains information obtained from authentic and highly regarded sources. Reprinted material is quoted with permission, and sources are indicated. A wide variety of references are listed. Reasonable efforts have been made to publish reliable data and information, but the author and the publisher cannot assume responsibility for the validity of all materials or for the consequences of their use.

No part of this book may be reprinted, reproduced, transmitted, or utilized in any form by any electronic, mechanical, or other means, now known or hereafter invented, including photocopying, microfilming, and recording, or in any information storage or retrieval system, without written permission from the publishers.

For permission to photocopy or use material electronically from this work, please access www.copyright.com (http://www.copyright.com/) or contact the Copyright Clearance Center, Inc. (CCC) 222 Rosewood Drive, Danvers, MA 01923, 978-750-8400. CCC is a not-for-profit organization that provides licenses and registration for a variety of users. For organizations that have been granted a photocopy license by the CCC, a separate system of payment has been arranged.

Trademark Notice: Product or corporate names may be trademarks or registered trademarks, and are used only for identification and explanation without intent to infringe.

Visit the Taylor & Francis Web site at
http://www.taylorandfrancis.com

and the CRC Press Web site at
http://www.crcpress.com

Dedication

Dr. Shigeo Shionoya 1923–2001

This volume is a testament to the many contributions Dr. Shionoya made to phosphor art and is dedicated to his memory.

In Memoriam

Akira Tomonaga
Formerly of Mitsubishi Electric Corp.
Amagasaki, Japan

The Editors

William M. Yen obtained his B.S. degree from the University of Redlands, Redlands, California in 1956 and his Ph.D. (physics) from Washington University in St. Louis in 1962. He served from 1962–65 as a Research Associate at Stanford University under the tutelage of Professor A.L. Schawlow, following which he accepted an assistant professorship at the University of Wisconsin-Madison. He was promoted to full professorship in 1972 and retired from this position in 1990 to assume the Graham Perdue Chair in Physics at the University of Georgia-Athens.

Dr. Yen has been the recipient of a J.S. Guggenheim Fellowship (1979–80), of an A. von Humboldt Senior U.S. Scientist Award (1985, 1990), and of a Senior Fulbright to Australia (1995). He was recently awarded the Lamar Dodd Creative Research Award by the University of Georgia Research Foundation. He is the recipient of the ICL Prize for Luminescence Research awarded in Beijing in August 2005. He has been appointed to visiting professorships at numerous institutions including the University of Tokyo, the University of Paris (Orsay), and the Australian National University. He was named the first Edwin T. Jaynes Visiting Professor by Washington University in 2004 and has been appointed to an affiliated research professorship at the University of Hawaii (Manoa). He is also an honorary professor at the University San Antonio de Abad in Cusco, Peru and of the Northern Jiatong University, Beijing, China. He has been on the technical staff of Bell Labs (1966) and of the Livermore Laser Fusion Effort (1974–76).

Dr. Yen has been elected to fellowship in the American Physical Society, the Optical Society of America, the American Association for the Advancement of Science and by the U.S. Electrochemical Society.

Professor Shionoya was born on April 30, 1923, in the Hongo area of Tokyo, Japan and passed away in October 2001. He received his baccalaureate in applied chemistry from the faculty of engineering, University of Tokyo, in 1945. He served as a research associate at the University of Tokyo until he moved to the department of electrochemistry, Yokohama National University as an associate professor in 1951. From 1957 to 1959, he was appointed to a visiting position in Professor H.P. Kallman's group in the physics department of New York University. While there, he was awarded a doctorate in engineering from the University of Tokyo in 1958 for work related to the industrial development of solid-state inorganic phosphor materials. In 1959, he joined the Institute for Solid State Physics (ISSP, Busseiken) of the University of Tokyo as an associate professor; he was promoted to full professorship in the Optical Properties Division of the ISSP in 1967. Following a reorganization of ISSP in 1980, he was named head of the High Power Laser Group of the Division of Solid State under Extreme Conditions. He retired from the post in 1984 with the title of emeritus professor. He helped in the establishment of the Tokyo Engineering University in 1986 and served in the administration and as a professor of Physics. On his retirement from the Tokyo Engineering University in 1994, he was also named emeritus professor in that institution.

During his career, he published more than two hundred scientific papers and authored or edited a number of books—the *Handbook on Optical Properties of Solids* (in Japanese, 1984) and the *Phosphor Handbook* (1998).

Professor Shionoya has been recognized for his many contributions to phosphor art. In 1977, he won the Nishina Award for his research on high-density excitation effects in semiconductors using picosecond spectroscopy. He was recognized by the Electrochemical Society in 1979 for his contributions to advances in phosphor research. Finally, in 1984 he was the first recipient of the ICL Prize for Luminescence Research.

Hajime Yamamoto received his B.S. and Ph.D. degrees in applied chemistry from the University of Tokyo in 1962 and 1967. His Ph.D. work was performed at the Institute for Solid State Physics under late Professors Shohji Makishima and Shigeo Shionoya on spectroscopy of rare earth ions in solids. Soon after graduation he joined Central Research Laboratory, Hitachi Ltd., where he worked mainly on phosphors and p-type ZnSe thin films. From 1971 to 1972, he was a visiting fellow at Professor Donald S. McClure's laboratory, Department of Chemistry, Princeton University. In 1991, he retired from Hitachi Ltd. and moved to Tokyo University of Technology as a professor of the faculty of engineering. Since 2003, he has been a professor at the School of Bionics of the same university.

Dr. Yamamoto serves as a chairperson of the Phosphor Research Society and is an organizing committee member of the Workshop on EL Displays, LEDs and Phosphors, International Display Workshops. He was one of the recipients of Tanahashi Memorial Award of the Japanese Electrochemical Society in 1988, and the Phosphor Award of the Phosphor Research Society in 2000 and 2005.

Preface

This volume originated from the *Phosphor Handbook* which has enjoyed a moderate amount of sale success as part of the CRC Laser and Optical Science and Technology Series and which recently went into its second edition. The original *Handbook* was published in Japanese in 1987 through an effort of the Phosphor Research Society of Japan. The late professor Shionoya was largely instrumental in getting us involved in the translation and publication of the English version. Since the English publication in 1998, the *Handbook* has gained wide acceptance by the technical community as a central reference on the basic properties as well as the applied and practical aspects of phosphor materials.

As we had expected, advances in the display and information technologies continue to consume and demand phosphor materials which are more efficient and more targeted to specific uses. These continuing changes in the demand necessitated an update and revision of the *Handbook* and resulted in the publication of the second edition which incorporates almost all additional topics, especially those of current interest such as quantum cutting and LED white lighting phosphor materials.

At the same time, it has also become apparent to some of us that the evolution of recent technologies will continue to place demands on the phosphor art and that research activity in the understanding and development of new phosphor materials will continue to experience increases. For this reason, it has been decided by CRC Press that a series of titles dedicated to Phosphor Properties be inaugurated through the publication of correlated sections of the *Phosphor Handbook* into three separate volumes. Volume I deals with the fundamental properties of luminescence as applied to solid state phosphor materials; the second volume includes the description of the synthesis and optical properties of phosphors used in different applications while the third addresses experimental methods for phosphor evaluation. The division of the Handbook into these sections, will allow us as editors to maintain the currency and timeliness of the volumes by updating only the section(s) which necessitate it.

We hope that this new organization of a technical series continues to serve the purpose of serving as a general reference to all aspects of phosphor properties and applications and as a starting point for further advances and developments in the phosphor art.

William M. Yen
Athens, GA, USA
October, 2006

Hajime Yamamoto
Tokyo, Japan
October, 2006

Contributors

Sohachiro Hayakawa
Formerly of The Polytechnic
  University
Kanagawa, Japan

Sueko Kanaya
Kanazawa Institute of
  Technology
Ishikawa, Japan

Yoshiharu Komine
Formerly of Mitsubishi Electric
  Corp.
Amagasaki, Japan

Kohei Narisada
Formerly of Matsushita Electric
  Ind. Co., Ltd.
Osaka, Japan

Kazuo Narita
Formerly of Toshiba Research
  Consulting Corp.
Kawasaki, Japan

Masataka Ogawa
Sony Electronics Inc.
San Jose, California

Shinkichi Tanimizu
Formerly of Hitachi, Ltd.
Tokyo, Japan

Koichi Urabe
Formerly of Hitachi, Ltd.
Tokyo, Japan

Taisuke Yoshioka
Formerly of Aiwa Co., Ltd.
Tokyo, Japan

Contents

Chapter 1		Measurements of luminescence properties of phosphors	1
	1.1	Luminescence and excitation spectra	2
	1.2	Reflection and absorption spectra	25
	1.3	Transient characteristics of luminescence	30
	1.4	Luminescence efficiency	34
	1.5	Data processing	38
	1.6	Measurements in the vacuum-ultraviolet region	45
Chapter 2		Measurements of powder characteristics	51
	2.1	Particle size and its measurement	52
	2.2	Methods for measuring particle size	57
	2.3	Measurements of packing and flow	74
Chapter 3		Optical properties of powder layers	79
	3.1	Kubelka-Munk's theory	79
	3.2	Johnson's theory	93
	3.3	Monte Carlo method	105
Chapter 4		Color vision	115
	4.1	Color vision and the eye	116
	4.2	Light and color	118
	4.3	Models of color vision	119
	4.4	Specification of colors and the color systems	121
	4.5	The color of light and color temperature	128
	4.6	Color rendering	130
	4.7	Other chromatic phenomena	132
Index			135

chapter one — sections one–five

Measurements of luminescence properties of phosphors

Taisuke Yoshioka and Masataka Ogawa

Contents

1.1 Luminescence and excitation spectra 2
    1.1.1 Principles of measurement 2
    1.1.2 Measurement apparatus 3
        1.1.2.1 Monochromator 3
        1.1.2.2 Light detector 7
        1.1.2.3 Signal amplification and processing apparatus 14
    1.1.3 Excitation source 16
        1.1.3.1 Ultraviolet and visible light sources 17
        1.1.3.2 Electron-beam excitation 22
    1.1.4 Some practical suggestions on luminescence measurements 25
1.2 Reflection and absorption spectra 25
    1.2.1 Principles of measurement 25
    1.2.2 Measurement apparatus 27
1.3 Transient characteristics of luminescence 30
    1.3.1 Principles of measurement 30
    1.3.2 Experimental apparatus 31
        1.3.2.1 Detector 31
        1.3.2.2 Signal amplification and processing 32
        1.3.2.3 Pulse excitation source 34
1.4 Luminescence efficiency 34
    1.4.1 Principles of measurement 34
    1.4.2 Measurement apparatus 35
        1.4.2.1 Ultraviolet excitation 35
        1.4.2.2 Electron-beam excitation 37
1.5 Data processing 38
    1.5.1 Spectral sensitivity correction 38
    1.5.2 Baseline correction 40
    1.5.3 Improvement of signal-to-noise ratio 40
Appendix 41
References 43

The luminescence properties of a phosphor can be characterized by its emission spectrum, brightness, and decay time. The absorption and reflectance spectra of phosphors provide additional information pertaining to both the basic luminescence mechanisms and their practical application.

This chapter describes how to measure the optical properties of a phosphor. The apparatus used and the method of measurement are introduced. The operating principles of each instrument employed are explained. Methods of obtaining meaningful data from raw experimental data are given.

1.1 Luminescence and excitation spectra

1.1.1 Principles of measurement

The luminescence spectrum is obtained by plotting the relationship between the wavelength and the intensity of the emitted light from a sample excited by an appropriate excitation source of constant energy. The excitation source can be light, an electron beam, heat, X-rays, or radiation from radioactive materials. The spectrum is obtained using a monochromator (see 1.1.2) equipped with an appropriate light detector.

In the case of an excitation spectrum, on the other hand, the relationship is obtained by observing changes in the emitted light intensity at a set wavelength while varying the excitation energy. When the excitation source is light, single-frequency light produced by a monochromator impinges on the sample and the emitted light intensity is recorded as the excitation wavelength is varied.

In a spectrum, light intensity at a given wavelength is expressed along the ordinate and the wavelength along the abscissa. The units of the ordinate are either irradiance E (W·m^{-2}) or number of photons E_p (photons·m^{-2}). The units of the abscissa are expressed in terms of wavelength λ (nm) or wave number $\tilde{v}$ (cm^{-1}).

Using these units, the spectrum irradiance is expressed as:

$$E(\lambda) = \frac{dE}{d\lambda} \quad \left(W \cdot m^{-2} \cdot nm^{-1}\right) \tag{1}$$

or

$$E(\tilde{v}) = \frac{dE}{d\tilde{v}} \quad \left(W \cdot m^{-2} \cdot \left(cm^{-1}\right)^{-1}\right) \tag{2}$$

and the spectral photon irradiance is expressed as:

$$E_p(\lambda) = \frac{dE_p}{d\lambda} \quad \left(photons \cdot m^{-2} \cdot nm^{-1}\right) \tag{3}$$

or

$$E_p(\tilde{v}) = \frac{dE_p}{d\tilde{v}} \quad \left(photons \cdot m^{-2} \cdot \left(cm^{-1}\right)^{-1}\right) \tag{4}$$

Chapter one: Measurements of luminescence properties of phosphors

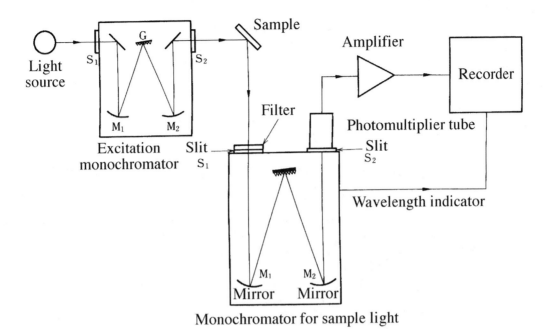

Figure 1 Spectroscopic measurement apparatus.

The units used depend on the purpose of the experiment. For energy efficiency, irradiance is employed and for quantum efficiency, photon irradiance is employed. The luminosity of a phosphor is expressed in terms of irradiance, which is obtained by integrating the spectral data, $E(\lambda)$, multiplied by the relative photopic spectral luminous efficiency, $V(\lambda)$, divided by the light equivalence value,[1,2] $K_m = 673$ lm·Watt^{-1}; that is,

$$L = K_m \int_0^\infty V(\lambda)E(\lambda)d\lambda, \quad \text{lm} \cdot \text{m}^{-2} \tag{5}$$

1.1.2 Measurement apparatus

The apparatus for measuring the spectral characteristics of phosphors is shown in Figure 1. The excitation source consists of the light source and a monochromator, which selects a specific wavelength range from the incoming light. (The monochromator can be replaced by a filter.) The light emitted from the sample is analyzed by a monochromator equipped with a light detector. The light detector transforms the photons into electrical signals. After the signals are amplified, they are recorded, typically on a strip chart recorder. It is often convenient to collect all spectral data in the form of digitized electrical signals and to use a computer for further processing the data.

The equipment used to measure the fluorescence characteristics of phosphors is as follows.

1.1.2.1 Monochromator

The monochromator is an apparatus used to select a particular wavelength of light. The monochromator consists of an entrance slit, an exit slit, a dispersing element for polychromatic light, and optics that focus the entrance slit image onto the exit slit. Monochromators can be classified by the dispersing element employed as either prism type, diffraction-

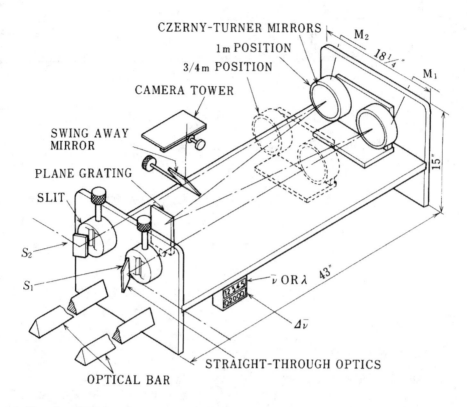

Figure 2 Interior of Czerny-Turner mount grating spectrometer. An illustration of Model 1702 of Jobin-Yvon-Spex is chosen because of its simplicity. This model has been replaced by Model 750M. (From *Model 1702 Instruction Manual*, Jobin Yvon-Spex, Edison, New Jersey. With permission.)

grating type, or interference type, the most popular being the latter two. The interference type is mainly used to measure light in the infrared region because these instruments are fast and have high sensitivity. The prism-type monochromator has the advantage of durability, and can be made very compact. The prism type, however, has lower resolution than the other two types, and is quite often employed as an optical filter for a diffraction-grating monochromator in high-resolution work.

The Czerny-Turner mount shown in Figure 1 is a typical diffraction-grating monochromator configuration.[3] The detailed internal structure can be seen in Figure 2.[4]

The optics of the Czerny-Turner instrument consist of two concave mirrors with equal focal length. The collimator mirror M_1 is positioned at a distance equal to its focal length from the entrance slit S_1. Light entering through the entrance slit is thus collimated and the parallel beam is diffracted by diffraction grating G. The diffracted beam is now monochromatic and is focused by the second concave mirror M_2 on the exit slit. The mirror M_2 is often called the camera mirror. By adjusting the angle of the diffraction grating with respect to the direction of the incident light, the wavelength of the diffracted monochromatic beam can be changed at the exit slit. One of the advantages of the Czerny-Turner mount is that the camera mirror cancels exactly the aberration generated by the collimator mirror, because the configuration is totally symmetric. Also, because the light is incident along the plane perpendicular to the grooves of the diffraction grating, the wavelength dependence on light polarization is small. Other than the Czerney-Turner mount, there are the Ebert-Fastie mount, which has a simpler optical configuration, and the double-grating mount for high-resolution studies.

The brightness of a monochromator is described by the aperture ratio F, defined as the focal length of the concave mirror, f, divided by its effective diameter D:

$$F = \frac{f}{D} \tag{6}$$

The ability to separate two closely spaced spectral lines is expressed in terms of a linear dispersion by:

$$\frac{dx}{d\lambda} = f \frac{d\theta}{d\lambda} \quad (\text{mm} \cdot \text{nm}^{-1}) \tag{7}$$

where two spectral lines separated by $d\lambda$ in wavelength at the exit slit are separated by a spatial distance of dx. The reciprocal linear dispersion is more commonly used, however:

$$\frac{d\lambda}{dx} = \frac{1}{f} \frac{d\lambda}{d\theta} \quad (\text{nm} \cdot \text{mm}^{-1}) \tag{8}$$

The product of the reciprocal linear dispersion and the slit width gives the separation of the two spectral lines (full width at half maximum) at a given wavelength. For observation of low light intensity, a monochromator having a small F number, i.e., having a short focal length, should be used. For high-resolution studies, on the other hand, a monochromator with a longer focal length is generally employed.

Figure 3 shows a microscopic cross-section of an echelette-type planar diffraction grating. As is seen in this figure, the grating has a saw-tooth shape. The angle between the grating plane and the saw-tooth plane is called the blaze angle θ. The grating constant d is defined as the pitch of the grooves. Light diffraction occurs when the following geometrical relation is satisfied:

$$m\lambda = d(\sin\alpha + \sin\beta), \quad (m = 0, \pm 1, \pm 2, \ldots), \tag{9}$$

where α is the angle between the line perpendicular to the plane of the grating and the direction of the incident light, and β is the angle between the same normal and the direction of the diffracted light. Light from adjacent grooves interferes constructively when a multiple, m, of the diffracted light wavelength is equal to the difference in the incident and diffracted light path lengths. This integer m is called the spectral order.

Geometrical optics describe the reflection condition when the incident angle, the diffraction angle, and the blaze angle satisfy the following relation:

$$\theta = \frac{\alpha + \beta}{2} \tag{10}$$

Under this condition, Eq. 9 becomes:

$$m\lambda = 2d \sin\theta \cos\frac{\beta - \alpha}{2} \tag{11}$$

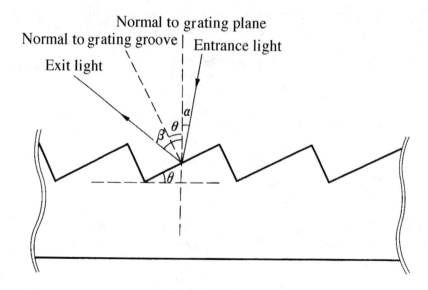

Figure 3 Cross-section of echelette plane grating.

When the incident light is perpendicular to the plane of the grating grooves, $\alpha = \beta = \theta$ holds and the following relation is obtained:

$$m\lambda = 2d \sin \theta \tag{12}$$

The first-order diffracted light is maximized at $\lambda = 2d \sin \theta$ and this λ is called the blaze wavelength.

The wavelength region where echelette gratings are most effectively utilized is defined by dividing the blaze wavelength by the grating order. For example, an echelette grating employed for the UV and visible regions has a 500-nm blaze wavelength; hence, it is useful between 250 and 750 nm in first order and between 125 and 375 nm in second order.

The term $\sin \alpha + \sin \beta$ in Eq. 9 expresses the degree of light dispersion, which can be seen to be proportional to m, λ, and d^{-1}. The resolution improves with longer wavelengths, with higher-order diffracted light, and with finer groove pitch of the grating.

The above situation applies for grating monochromators in general, but when dealing with the UV region, there is an additional consideration. Because air absorbs light of wavelengths below 200 nm, the optical path must be kept in vacuum. Since no materials with good reflectivity in this wavelength region are available, no reflecting mirrors are used. Because even window materials such as LiF do not transmit light below 105 nm, the light detector must also be placed in the vacuum. The vacuum of the sample chamber, which is located between the light source and the rest of the optical components, normally can be independently broken to change samples.

The optical configuration of a vacuum-UV spectrometer consists of a concave diffraction grating and entrance and exit slits. This configuration, proposed by Seya and Namioka,[5] is shown in Figure 4. The concave diffraction grating and the entrance and exit slits are placed on a Rowland circle. Only the grating rotates; the other components are fixed in position. The diameter of Rowland circle is defined by the center of the concave diffraction grating and its radius of curvature. When the entrance slit and exit slits are positioned so as to form an angle of 30°15' with a line from the center of the grating to the point immediately opposite on the Rowland circle, and the distance between the center

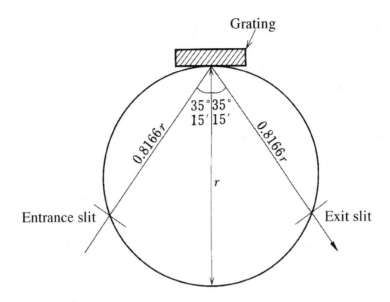

Figure 4 Configuration of optical components of Seya-Namioka vacuum ultraviolet spectrometer.

of the grating and the slits is 0.8166r, where r is the radius of curvature of the concave grating, as in Figure 4, the entrance slit focuses at the output with minimum distortion, which is caused by the rotation of the grating.[6]

In the wavelength region below 50 nm, light reflectivities become extremely small, and the total reflected light must be measured at a wide angle of incident light. In this optical configuration, since the two slits and the grating are on a Rowland circle, their geometrical relations become quite complex.

1.1.2.2 Light detector

Light is usually detected by converting its energy to electrical energy. The two light conversion elements most commonly used due to their reliability and ease of handling are photomultiplier tubes and solid-state detectors. There are a number of other methods of detecting light, for example, by using a thermoelectric element that measures the thermal energy generated by absorbed light energy or by observing the chemical products formed in a photochemical reaction.

Photomultiplier tube.[7] The photomultiplier tube is frequently used for detecting UV and visible light. Because the initial photoelectrons are multiplied many fold and because of their fast response time, photomultiplier tubes are employed for measuring very low-level light and fast transient phenomena. The inside structure of a side-on type photomultiplier tube is shown in Figure 5. As can be seen in this figure, the photomultiplier tube consists of a photoelectric surface (cathode) from which photoelectrons are generated by the incident photons. The photoelectrons then enter into a multistage dynode structure in which they are accelerated by the voltage applied to adjacent dynodes. Each dynode stage produces many secondary electrons so that the initial number of photoelectrons is multiplied many fold. All electrons generated in this way are collected by the anode. Anywhere from several tens to hundreds of volts are applied between the dynodes. Since the number of secondary electron emission surfaces in a photomultiplier tube K are commonly about ten and the electron multiplication factor δ, at each surface is a factor of 4 to 5, the overall multiplication δ^K is of the order of 10^6.

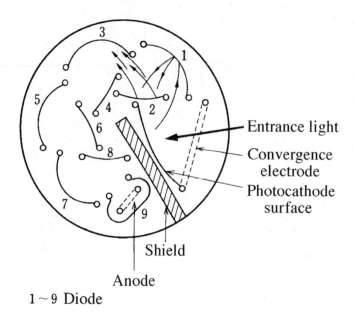

Figure 5 Side-on type photomultiplier tube.

A photomultiplier tube having the appropriate spectral sensitivity for the wavelength region of interest must be chosen. Figure 6 shows spectral sensitivity curves for different photoelectric surfaces.[6] For the UV and visible regions, the multi-alkaline metals (Na-K-Cs-Sb) exhibit the highest sensitivity. A Ga-As surface shows good sensitivity well into the near-infrared region. For studies in the longer wavelength region, an Ag-O-Cs surface is employed. For the vacuum-UV (VUV) region, a Cs-I and Cs-Te photoelectron surface is employed. Beyond this region, VUV light is converted into visible light by means of a phosphor screen using a phosphor such as sodium salicylate; the visible light is detected by a photomultiplier whose sensitivity is appropriate for the emitted light of the phosphor.

When an electric potential is applied to a photomultiplier tube, even in the absence of photons, a minute current flows through the tube. This current is called the dark current and is mainly due to thermal electron emission from the photoelectron surface. Besides this constant thermal emission, there is irregular shot noise caused by discharges in the residual gases and by light emission from the glass envelope caused by the bombardment of electrons. When measuring very low-level light, it is important to select an appropriate low dark-current photomultiplier tube and sometimes it is useful to cool the photomultiplier tube to minimize thermal noise.

One way to supply voltage to each dynode stage of a photomultiplier tube is shown in Figure 7. The value of the resistors R is set so that the maximum current at a given anode voltage is approximately ten times the anode current. When an AC signal is observed, the peak anode current can be unexpectedly large, so capacitors are provided in the later stages of dynodes to stabilize their voltage.

The following cautions must be taken when handling a photomultiplier tube:

1. Even when voltage is not applied to the tube, the tube should not be exposed to light.
2. A few hours prior to measurement, voltage should be applied to the tube in the dark to stabilize the output.

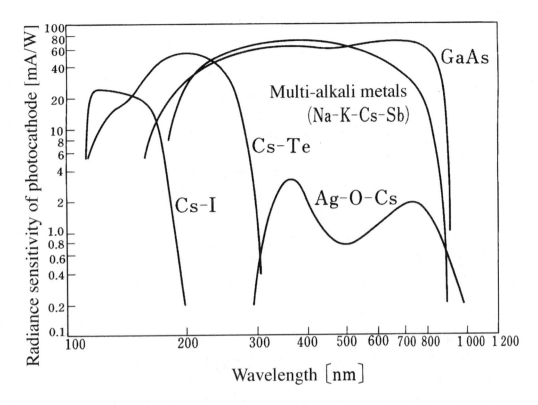

Figure 6 Spectral sensitivity curve of various photoelectric surfaces. (From *Photomultiplier Tubes Catalog*, Hamamatsu Photonics, Shizuoka, Japan, August 1995. With permission.)

3. Tubes must be operated under the manufacturer's specified conditions. Particularly, the anode current must not exceed one tenth of the maximum allowed anode current, except during transient peaks measurements.
4. Tubes are extremely delicate and great care must be taken in their handling.

Solid-state detectors. For *p-n* junctions in semiconductors, a region depleted of mobile charge carriers with a high internal electric field across it exists between the *p-* and *n*-type materials. This region is known as the depletion region. When light irradiates the depletion region, electron-hole pairs are generated through the absorption of photons and the internal field causes the electrons and holes to separate.

This accumulated charge can be detected by measuring the electric potential between the *p* and *n* regions while the device is open-circuit (the photovoltaic mode of operation). The charge can also be detected by measuring the current flow between the *p* and *n* regions by applying a reverse bias (the photoconductive mode of operation).

The most common semiconductor material used for photodiodes is silicon. A typical structure is shown in Figure 8.[8] It should be noted that electrical contact to the semiconductor material is always made via a metal-n^+ (or -p^+) junction. Silicon photodiodes have a bandgap of 1.14 eV, with quantum efficiencies up to 80% at wavelengths between 0.8 and 0.9 μm. Detection efficiency may be increased by providing antireflection coatings on the front surface of the detector consisting of a $\lambda/2$ coating of SiO_2.

The amount of dark current of solid-state detectors can be reduced by cooling the detectors, as is the case in photomultiplier tubes. For this reason, the signal-to-noise ratio of detectors can be improved by using thermoelectric devices to cool them.

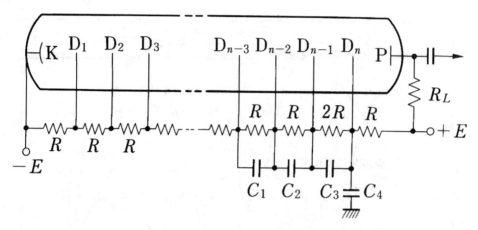

(a) AC measurement

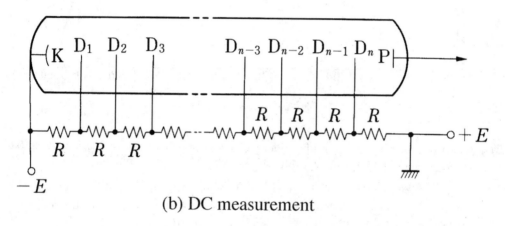

(b) DC measurement

Figure 7 Method of supplying voltage to photomultiplier tube.

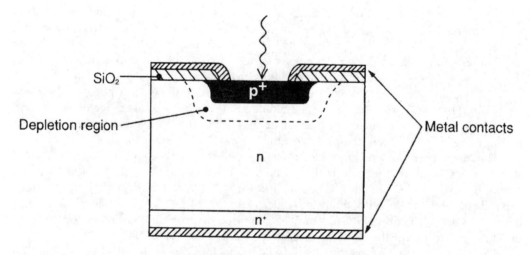

Figure 8 Silicon photodiode structure. (From Wilson, J. and Hawkes, J.F.B., *Optoelectronics, An Introduction*, Prentice-Hall, 1989, 284. With permission.)

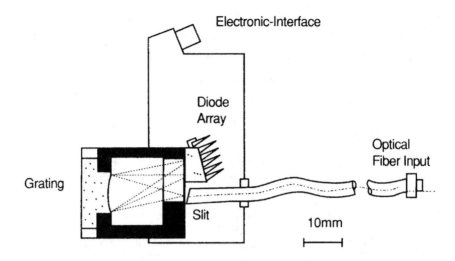

Figure 9 A compact spectrometer using a photodiode array. (From *Monolithic Miniature Spectrometer, Product Information*, Carl Zeiss, Germany. With permission.)

Single-channel and multichannel detectors. Light detectors employed for spectral studies can be classified as single-channel or multichannel detectors. The single-channel detector has a single light-detecting element. Typical examples of these detectors are the photomultiplier tube and the solid-state photodiode. The single-channel detector is placed in front of the exit slit of a monochromator. The detector measures light intensity at a given wavelength. A spectrum is obtained by scanning the wavelength range of interest, taking intensity data at each wavelength.

The multichannel detector has multiple light-detecting elements arranged linearly or in two dimensions, with each element operating individually. Examples of this type of detector are MOS-FET photodiode arrays and charge-coupled devices (CCDs). A classic example of a multichannel detector is a photographic plate used in conjunction with a spectrograph. The multichannel detector is positioned at the focal plane of the light exit of a monochromator with the exit slit removed. The detector can, therefore, cover a wide wavelength region. The width and height of each photodetecting element are equivalent to the width and height of the exit slit in a monochromator. When a multichannel detector is used, the monochromator does not have to scan the wavelength region of interest and it can measure the total spectrum within several to several hundreds of milliseconds. The other advantage of a multichannel-type detector is that since it accumulates the light energy, signal levels can be increased by extending the exposure time.

An efficient spectrum-measuring system can be built in this way, as the data can be read out electrically and digitized, then fed into a personal computer for further processing. Many such systems are commercially available, together with the appropriate software. An extremely compact spectrometer of this type is shown in Figure 9.[9]

Figure 10 shows an equivalent circuit for an all-solid-state one-dimensional array detector. The operating principle is that each MOS-FET photodiode is initially charged by an applied electric field. When light irradiates on the photodiode, electron-hole pairs are generated and the holes discharge the previously accumulated electric charge, so that the recharge of the diode is proportional to the light intensity. The solid-state detector feeds the recharging current by means of electronic switching. Namely, in order to scan the channel, a shift register provides gate signals to each FET in succession.

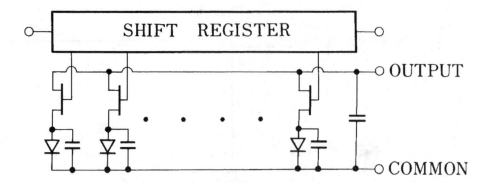

Figure 10 An equivalent circuit for an all solid-state one-dimensional array detector.

The standard photodiode array element is 25 µm wide and 2.5 mm high. The photodiode array itself is 25.6 mm wide and contains 1024 diodes.

Charge-coupled device (CCD). A spectral image formed on the array detector is converted into an electrical signal by each discrete element. If the detector elements are arranged in two dimensions and their number is typically 380,000, reading the electrical signal in each element requires a special technique. One method of obtaining the electrical signal sequentially is to use a charge-coupled device (CCD).

The upper portion of Figure 11 illustrates the basic configuration of the CCD consisting of a metal-oxide-semiconductor (MOS) capacitor.[10] A layer of silicon dioxide is grown on a *p*-type silicon substrate; a metal electrode is then evaporated on the oxide layer. The metal electrode acts as a gate and is biased positively with respect to the silicon. Electron-hole pairs are formed when the device is irradiated by light and electrons are attracted and held at the surface of the silicon under the gate when the voltage is positive. The electrons are effectively trapped within a potential well formed under the gate contact. The amount of charge trapped in the well is proportional to the total light flux falling onto the device during the measurement period.

The lower portion of Figure 11 illustrates how the trapped electrons are sequentially read out. The gate potentials are supplied by three voltage lines (L_1, L_2, L_3), which are connected to every third electrode (G_1, G_2, G_3) as shown. If the potential of L_1 is positive V_g, while L_2 and L_3 are at zero potential, a photogenerated charge proportional to the light falling on G_1 will be trapped under the electrode. (Figure 11(a)). After a suitable integration time, the charge can be removed by applying a voltage V_g to L_2 while maintaining L_1 at V_g; the charge initially under L_1 will now be shared between G_1 and G_2 (Figure 11(b)). If the potential of L_1 is then reduced to zero, all the charge that was initially under G_1 is moved to G_2 (Figure 11(c)). Repetition of this cycle will progressively move the charge along the MOS capacitors from left to right. At the end of the line, the amount of charge arriving as a function of time then provides a sequential scan of the "G_1" detector output.

In order to achieve a faster scanning rate, a second CCD array (the transport register) is provided. The transport register is shielded from the incident light and lies alongside the light-sensing array. Once a charge has built up in the sensing array, it is transferred "sideways" to the transport register and can be read out at the output of the transport register. This readout can take place at the same time as a new image is being built up in the sensing array.

Chapter one: Measurements of luminescence properties of phosphors 13

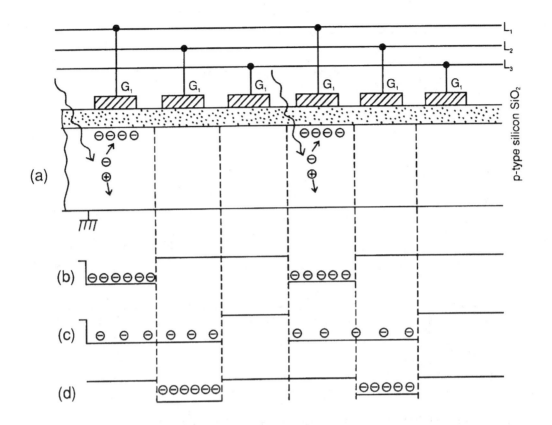

Figure 11 Basic CCD array composed of a line of MOS capacitors. (From Wilson, J. and Hawkes, J.F.B., *Optoelectronics, An Introduction*, Prentice-Hall, 1989, 296. With permission.)

Two-dimensional arrays based on the above one-dimensional designs are also possible and are known as frame-transfer devices. The transfer registers feed into a readout register running down the edge of the device. The contents of each line are read out in sequence into the readout register so that the signal appearing at the output of these registers represents a line-by-line scan of the image.

A great advantage of the two-dimensional CCD detector is that by introducing multiple images on different portions of the CCD, the upper, middle, and lower portions for example, separate spectra can be obtained simultaneously by reading out the sectional data separately.

The MOS-CCD has a quantum efficiency of about 45 to 50% at the peak of its sensitivity, 750 nm. The quantum efficiency can be improved at shorter wavelengths by coating the elements with a fluorescent dye that converts UV light to longer wavelengths to match the maximum quantum efficiency of the MOS photodetector. The efficiency of the detector in the longer wavelength region can be improved by making the potential well of the depletion region deeper than that of a standard chip. Another technique to improve the efficiency of the CCD device is to make the substrate very thin. In this back-thinned CCD, light is incident on the back rather than the front. This is because the gates are on the top, creating a thick layer for electrons to travel through and thus reducing their probability of reaching the depletion region. With the back-thinned chip, the chances for an electron to reach the depletion region are greater, and thus the quantum efficiency is higher. The spectral response curves of a variety of CCDs are shown in Figure 12.[11,12]

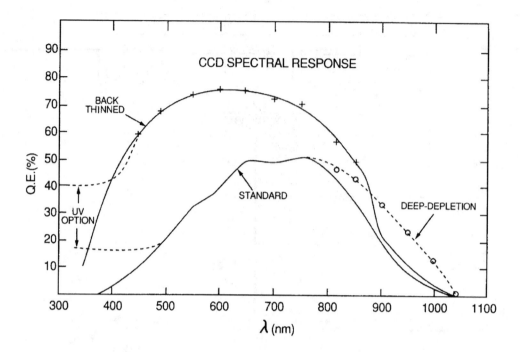

Figure 12 Spectral response curves of a variety of CCD detectors. (From *Guide for Spectroscopy*, Jobin Yvon-Spex, Edison, New Jersey, 1994, 217. With permission.)

The noise in a CCD is composed of shot noise, dark current, and read-out noise. CCDs can be cooled either thermoelectrically or with liquid nitrogen to reduce the dark current and associated thermal noise. The liquid nitrogen-cooled CCD is one of the most sensitive detectors available, having a dark signal of 1 electron per pixel per hour. The photodiode array device previously described cannot be cooled to liquid nitrogen temperature as it must have associated electronic circuits that cannot operate at low temperatures.

CCDs come in standard sizes of 1152 × 298, 512 × 512, and 578 × 385 pixels, with individual pixel sizes of 22 × 22 μm.

Image intensifier.[13,14] To measure extremely weak light, a detector with an image intensifier is used. Intensifiers are particularly useful when used in conjunction with solid-state detector arrays, as the latter are not very sensitive to low light levels, relative to photomultipliers. The operating principle of the image intensifier is shown in Figure 13 and is similar to that of a photomultiplier tube. When light is incident on the photocathode, photoelectrons are generated. The photoelectrons travel through a microchannel plate to the phosphor screen, being accelerated by a potential applied between the photocathode and the phosphor screen. The microchannel plate consists of thin metalized glass fibers. The electrons from the photocathode collide along the metalized walls, generating secondary electrons. Thus, multiplication by more than a factor of 1000 can be obtained. The photocathode material employed will vary depending on the wavelength sensitivity required, as in the case of a photomultiplier tube.

1.1.2.3 Signal amplification and processing apparatus

The signal from the photodetector must be further processed electrically to obtain meaningful data. In order to acquire data with a good signal-to-noise ratio, a variety of techniques are employed. In this section, some useful techniques are described.

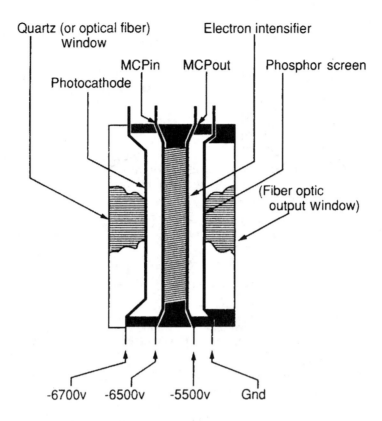

Figure 13 Inside structure of an image intensifier. (From *Applications of Multi-channel Detectors Highlighting CCSs*, Jobin Yvon-Spex, Edison, New Jersey. With permission.)

DC current-voltage converter. The output signal from photoelectric detectors such as a photomultiplier tube or a photodiode is in the form of a photocurrent. To display the signal on a strip-chart recorder, the photocurrent must be converted into a voltage. A current-voltage converter is used for this purpose. The simplest converter is a load resistor placed serially between the output, such as the photomultiplier anode plate, and the ground to allow observation of the output voltage. To observe this voltage directly by using a measuring apparatus (e.g., a strip-chart recorder), the instrument must have a higher input impedance than the load resistance. For DC measurements, an impedance conversion circuit as shown in Figure 14 is frequently used. An operational amplifier that requires an input off-set current much smaller than the photocurrent can be used as a DC amplifier with a V/A conversion ratio of up to 10^9.

Lock-in amplifier. The circuit diagram for this type of amplifier is shown in Figure 15. When the light signal is chopped at a certain frequency, the detector output consists of the signal and the non-signal component, alternately. The modified signal passes through a coupling capacitor and only the chopped frequency component is amplified by the synchronous amplifier. Using a reference signal generated by the chopper, the signal is phase-detected relative to the modulated signal. A phase shifter is adjusted to give a maximum output signal and a low-pass filter is adjusted for a time constant that optimizes the signal-to-noise ratio. Since only the input signal's phase is the same as that of the reference, the stray light that did not pass through the light chopper and other random electric noises are eliminated.

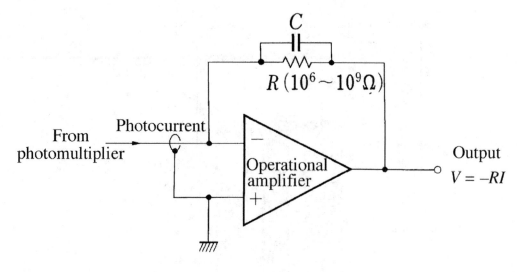

Figure 14 Current-voltage converter employing an operational amplifier.

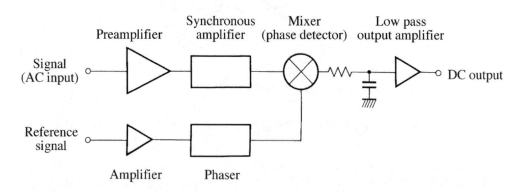

Figure 15 Block diagram of a lock-in amplifier.

Photon counter. A photon counting technique is employed when the light signal is extremely weak. The output from the photomultiplier can be observed as discrete photoelectron pulses. The intensity of the light signal is proportional to the number of photoelectron pulses per unit time. Pulses with a range of amplitude are input to a pulse-height discriminator circuit that distinguishes the signal from the dark current, as shown in Figure 16. In order to eliminate contributions from stray light and other noise sources from the signal, a light chopper is employed, as in the case of the lock-in amplification technique. The output signal from the detector contains [signal + stray light + noise] and [stray light + noise] on alternate half cycles so the difference yields the signal only. This technique is particularly useful to reduce shot-type noise because its occurrence is random, it contributes to both cases.

1.1.3 Excitation source

In order to observe fluorescence from a sample, some form of energy must be supplied to the sample. In this section, a variety of excitation sources for the investigation of fluorescence properties of material is described.

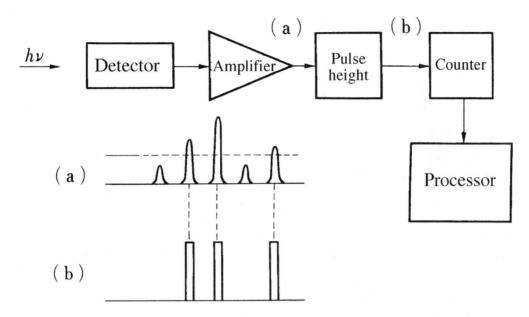

Figure 16 Block diagram and operational principle of photon counter. (a) Relation between input photoelectron pulse to the pulse-height discriminator and the discriminator voltage. (b) Output pulse to be counted from the pulse-height discriminator corresponding to the input signal.

1.1.3.1 *Ultraviolet and visible light sources*

The photo-excited fluorescent spectrum of material is most commonly studied in the ultraviolet to visible region. An appropriate light source is selected for experimental purposes in combination with a suitable filter and/or monochromator. The usual sources are discussed below.

Tungsten lamp. A tungsten lamp is easy to handle, is economical, has a relatively long life, and exhibits radiation characteristics similar to those of black-body radiation. As the temperature of the filament is raised, the radiation intensity in the short-wavelength region increases. Using a quartz or borosilicate glass envelope, which has good transmittance in the short-wavelength region, this lamp is useful for a variety of optical measurements from the near-UV to the near-IR. One disadvantage of tungsten lamps is that during operation, tungsten evaporates from the filament and is gradually deposited on the inside wall; this causes blackening of the surface of the envelope and absorption of light at shorter wavelength. In order to avoid the deposition of tungsten, either a sufficiently large envelope is used or Ar gas is introduced into the envelope.

To increase the life and improve the stability of the lamp, a metal halogen lamp has been developed. The metal halogen lamp contains small amounts of halogen gases such as bromine or iodine. Tungsten vapor from the high-temperature filament reacts with halogen in the vicinity of the lower-temperature wall and becomes a volatile tungsten halogenide. The tungsten halogenide is carried to the high-temperature filament by convection and decomposes into tungsten and halogen. The tungsten is redeposited on the filament. Wall blackening and tungsten loss from the filament can be avoided by this process. Metal halogen lamps can thus be operated at higher temperatures than conventional tungsten lamps, resulting in increases in radiation intensity in the short-wavelength region, in a doubling of lamp life, and in a higher lamp efficiency.

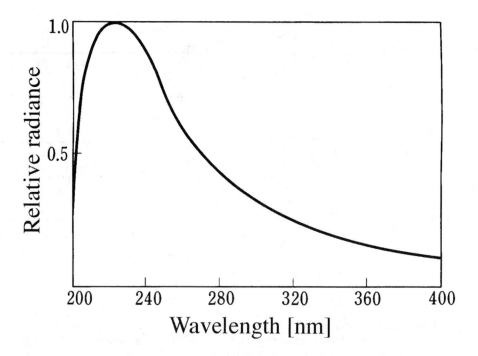

Figure 17 Spectral intensity distribution of hydrogen discharge lamp.

Discharge lamp. Discharge lamps most often used as excitation sources are hydrogen (deuterium), xenon, and mercury lamps.

Hydrogen discharge lamp. The hydrogen discharge lamp contains several torr pressure of either hydrogen or deuterium gas. The lamp is operated with a DC discharge between the hot electrodes. As is shown in Figure 17, the emission spectrum of the lamp is continuous in the ultraviolet region. High-intensity lamps are equipped with a glass jacket in which cooling water is circulated. The jacket encloses the entire lamp except around the window area. The window is made of quartz, which has high UV transmission.

Xenon discharge lamp. This lamp emits high-intensity light from the ultraviolet through the visible and infrared regions. Its relatively continuous spectrum is shown in Figure 18. Two types of this lamp are available: one in which the electrode gap is short (2 to 10 mm) with the gas at high pressure (several tens of atmospheric pressure), and one in which the electrode gap is long (several tens of cm) with low gas pressure. The light-emitting portion in the short-gap lamp is concentrated in the vicinity of the cathode area, so that this lamp can be regarded as a point source. The position of this bright discharge point tends to fluctuate, however, so caution must be taken when the source is focused on a sample. The long-arc lamp emits lower intensity light than the short-arc lamp. It emits a stable light output and is used as a standard in the UV region.

Mercury discharge lamp. The light emitted from a mercury discharge lamp spans the wavelength region from 185 to 365 nm. This lamp is the most common light source in the ultraviolet region. The mercury vapor pressure in the lamp is anywhere from below 1 mmHg to 50–200 atm, depending on the operating temperature. The spectrum of the output changes as the mercury pressure changes. At higher mercury pressures, the main line emissions broaden and their wavelength shifts toward longer wavelength; a continuous emission component also appears. At high mercury vapor pressures, the radiation

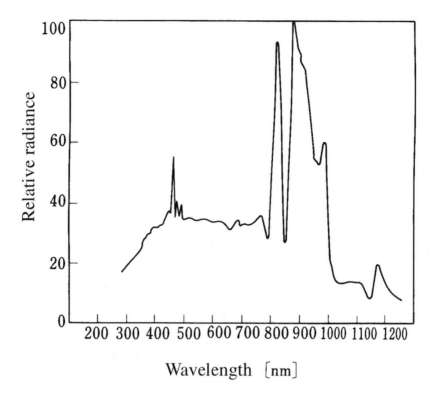

Figure 18 Emission spectrum of xenon short-arc lamp.

at 253.7 nm disappears due to self-absorption; the 365-nm line then becomes the main contributor to the UV region. A typical emission spectrum of the lamp is shown in Figure 19. As can be seen from the spectrum, the emissions are concentrated at particular wavelengths. Taking advantage of the nature of the spectrum, a strong monochromatic light source can be obtained by choosing particular wavelengths.

In order to eliminate the visible output of the mercury emission, a colored glass filter is employed such as the Toshiba UV-D33S or Corning 7-37 filters. For further absorption of visible light, a saturated aqueous solution of nickel sulfate is used for isolating 254-nm light and copper sulfate for isolating 365-nm light.

There are three kinds of commercially available mercury discharge lamps:

- A low-pressure mercury lamp in which the temperature of the lamp wall is relatively low and the main emission is at 254 nm. The typical input power of the lamp is anywhere from 5 to 20 W. There are two kinds of luminescence-detection (black light) lamps, which combine a glass filter with a low-pressure mercury lamp: one is for short (254 nm) and the other is for longer wavelengths (365 nm). The latter uses a filter containing a UV-emitting phosphor.
- A medium-pressure mercury lamp. By raising the wall temperature, the intensity of the longer wavelength components can be increased. The lamp has intermediate characteristics between the low- and the high-pressure lamps and is operated at 100 to 200 W input power. The main emission wavelengths are 254, 313, and 365 nm.
- A high-pressure mercury lamp. The lamp wall temperature is more than 200°C and this lamp can be used as a high-intensity point source. The lamp is operated at 150 to 2000 W input power. Recently, a stable and high-intensity UV source for appli-

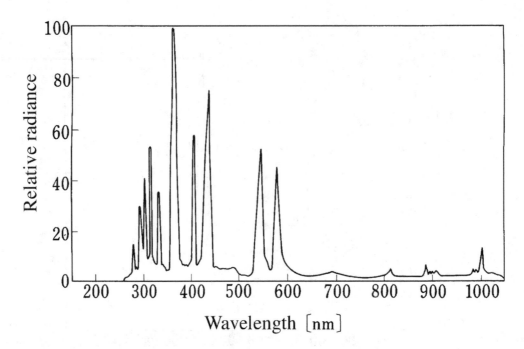

Figure 19 Emission spectrum of an ultra-high-pressure mercury lamp.

cation in photolithography has been developed; it contains a mixture of rare gas and mercury vapor.[15]

Laser.[16,17] A laser is an excellent monochromatic light source and has a radiative power at a given frequency several orders of magnitude greater than that of other light sources. Some lasers can operate in a pulsed mode and produce extremely short pulses. An appropriate gas, solid-state, liquid or dye laser, or semiconductor laser can be chosen depending on the experimental requirements. According to the mode of operation, lasers are either operated in continuous wave (cw) mode or in a pulsed mode. In some lasers, the output wavelength is tunable in a limited range. In the following, lasers useful for measuring luminescence properties are described.

Typical gas lasers used for the study of luminescence are the He-Ne, Ar$^+$ ion, Kr$^+$ ion, He-Cd, N$_2$, and excimer lasers.

- The He-Ne laser produces lasing line emission at 632.8 nm with high coherence, directionality, and wavelength stability. The output power can reach 35 mW. This laser is most frequently employed to align optical instruments.
- The Ar$^+$ ion laser has a total of ten lasing lines in the visible region 454 to 529 nm operating in cw mode. The most prominent line is at 514.5 nm, with an output power of up to 10 W. The total output power in the entire visible region can reach 25 W. This laser generates three lasing lines in the ultraviolet (wavelengths are 351.1, 351.4, and 363.8 nm), each with output powers of up to 1.5 W.
- The Kr$^+$ ion laser has a total of 15 lasing lines in the ultraviolet to visible region (350–676 nm) operating in cw mode. Unlike the Ar$^+$ laser, the Kr$^+$ laser has its strongest line in the red at 647.1 nm, with output powers up to 3.5 W.
- The He-Cd laser uses a mixture of He gas and Cd metal vapor, and has emission peaks in the ultraviolet and visible region. When it is operated in the cw mode, the

325-nm peak is most prominent, with output powers of 100 mW. This laser is very useful as an ultraviolet excitation source for measuring luminescence spectra.
- The N_2 laser is pulse-mode operated. There are high- and low-pressure types of N_2 lasers, differing in pulse width. The high-pressure N_2 laser has a pulse width of 0.1 to 1 ns, whereas the low-pressure N_2 laser has one of 5 to 10 ns. The N_2 laser has output peak power of as much as 200 kW and can be operated at a repetition rate of 100 Hz. The lasing wavelength of an N_2 laser is 337.1 nm. The laser is utilized as a high output power laser in the ultraviolet region, useful for high-intensity excitation of luminescence and for pumping dye lasers.
- The excimer laser. A series of excimer lasers is available. When the laser is operated in the pulse mode, all the excimer gases produce strong ultraviolet outputs: ArF excimer lases at 193 nm, KrF at 248 nm, XeCl at 308 nm, and XeF at 351 nm. The peak power of these laser is typically 20 MW (pulse energy of 200 mJ) but can produce 100 MW (1J) with pulse widths of 10 ns and repetition rates of 100 to 200 Hz. The lasers are useful for high-intensity excitation of luminescence and for pumping dye lasers.

The solid-state lasers employed for luminescence study are:

- The Nd^{3+}:YAG (yttrium aluminum garnet, $Y_3Al_5O_{12}$) laser is pumped by a xenon flash lamp, the most efficient lasing line being at 1.064 µm. Both continuous-mode lasing, which is now seldom used, and pulse-mode lasing can be obtained. The pulse-mode operation by Q-switching generates peak powers up to 110 MW (pulse energy of 1J) with a pulse width of 8 to 9 ns and repetition rates of 10 to 100 Hz. Since an extremely high output power can be achieved by this class of lasers, light can be generated at other frequencies using nonlinear processes. For example, the Nd^{3+}-YAG laser can generate (using the proper anharmonic crystal) light in the second order of 532 nm, in third order of 355 nm, and in fourth order of 266 nm, with conversion efficiencies of 30 to 40, 20, and 10%, respectively. Both the fundamental laser and its harmonics can be used to excite luminescence or to pump dye lasers. The laser can be mode-locked by acousto-optical modulation to yield extremely short pulse widths of 90 to 100 ps, with repetition rate of 100 MHz. The time-averaged power of these lasers can be as high as 7 W. This laser is useful for observing high-speed transient phenomena.
- The Nd^{3+}:YVO_4 laser has the same lasing line as Nd^{3+}:YAG at 1.064 µm. The crystal can be pumped by a laser diode (semiconductor laser) and can be operated in cw mode. Output power of the second harmonic (532 nm) has reached 5 W and this output is used to pump Ti^{3+}:sapphire lasers. A compact laser device can be constructed using laser diodes as a pumping source.
- The Ti^{3+}:sapphire (Al_2O_3) laser output is tunable and can be continuously varied from 670 to 1100 nm, in both cw-mode and pulse-mode operation. An 18% pumping power efficiency in cw-mode operation can be obtained when the crystal is pumped by the 532-nm laser output of a diode-pumped Nd^{3+}:YVO_4 laser, as noted above. The same pumping method is employed to attain mode-locked pulse operation. The mode-locking is carried out utilizing the Kerr effect.

The Ti^{3+}:sapphire laser gives very stable and extremely short pulses of 20 fs to 60 ps, tunable within the wavelength range of 680 to 1100 nm. The time- averaged output power of a commercially available laser is of the order of 2 W at 790 nm. Utilizing the second harmonic, light pulses at 395 nm with 150-mW power output have been obtained. By combining the fundamental and the second harmonics of

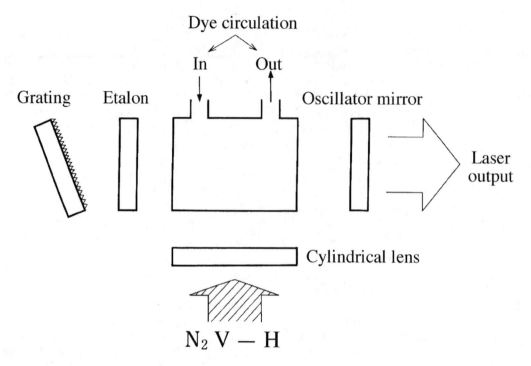

Figure 20 Configuration of a dye laser.

this laser, very useful femtosecond and picosecond laser pulses, tunable from the near-ultraviolet to near-infrared, can be obtained with a window around 600 nm.
- Dye laser. The lasing wavelength of a dye laser can be varied continuously from 370 to 1036 nm, depending on the dye employed.[11] As the fluorescence spectrum of a dye is, in general, a broad band, narrow-band lasing can be achieved using a diffraction grating or other wavelength-tuning element. By scanning the grating, the lasing wavelength can be changed. In order to obtain a higher output power, amplification stages can be used. The typical optical arrangement of a dye laser is shown in Figure 20.

Pulsed lasers can be used to pump dye lasers. Two approaches are available for obtaining output at a particular UV wavelength using dye lasers. One is to use an appropriate ultraviolet dye, and the other is to use the second or third harmonics of a visible dye laser. For example, the third harmonic of a visible dye laser pumped with a mode-locked Ti:sapphire laser can cover the wavelength region 273 to 322 nm with output powers of 10 to 120 mW.[18]

Other pumping sources include the N_2 laser, the XeCl excimer laser, and the Q-switched Nd^{3+}:YAG laser. Among dyes, the Rhodamine 6G laser covering a wavelength range of 565 to 620 nm is most efficient.

Excimer lamp. An excimer lamp containing a mixture of rare gas and halogen gas has been recently developed and commercialized by Ushio Co. The main output wavelengths are 172, 222, and 308 nm.[19]

1.1.3.2 Electron-beam excitation

Electron-beam energies used to excite phosphors range from several electron-Volts to several tens of keV. An electron beam of up to several hundred electron-Volts is called a

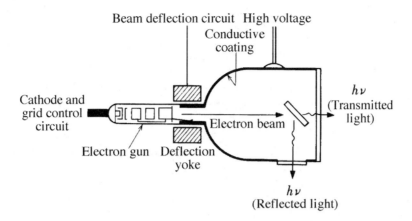

Figure 21 Structure of an electron-beam excitation apparatus.

low-energy beam, while beams of several kiloelectron-Volts energy are called high-energy beams. A low-energy electron beam penetrates into a phosphor particle a distance of only a few atomic layers, whereas a high-energy electron beam can excite the entire phosphor crystal.

Electron-beam excitation utilizes energy almost four times higher than optical excitation, which is several electron-Volts at most. In a cathode-ray tube, an electron beam of several tens of kiloVolts energy is converted to visible light by phosphor particles. The phosphor must be placed in a vacuum to allow excitation by the high-voltage electron beam. An example of the apparatus used for electron beam excitation is shown in Figure 21.

In order to observe reflected luminescence from a powder phosphor, the phosphor powder is tightly packed into the cavity in a metal sample holder. This cavity-filling technique is a simple and convenient way to test phosphors. The reflected luminescence is observed on the same side as the electron excitation; this method is different from that of the conventional cathode-ray tube and yields slightly different results. To simulate the configuration used to observe transmitted luminescence, a slide glass is coated with the sample phosphor by either sedimentation or slurry coating. Since phosphors are generally good insulators, the sample charges up as the electron beam is turned on; because of this space charge, the beam is deflected to areas where the electrical potential is highest and optical measurements can be disturbed. To avoid space charges, a tin-oxide-coated glass is employed, making the holder conductive. Another way to make a sample conductive is to evaporate aluminum onto the sedimented phosphor layer.

To prepare a sample for transmission luminescence measurements, the thickness of the phosphor layer is critical. If the phosphor layer is too thin, some of the electrons pass through the layer without colliding with the particles and thus do not fully excite the phosphor. On the other hand, if the phosphor layer is too thick, and since the accelerated electrons have a certain penetration depth, layers below this depth are not excited and a portion of the resulting luminescence can be self-absorbed. The luminescence spectrum therefore changes, depending on the sample thickness, and this thickness must be optimized to obtain the greatest possible light output. The optimum thickness depends on the particle size distribution and the excitation conditions, so the optimum layer thickness is best determined empirically. In the case of the measurement of reflected luminescence, the thickness is adjusted so that the sample does not charge up; hence, the place being irradiated does not change. Optically, this case can be regarded as being equivalent to measuring a sample with infinite thickness.

There are two ways to apply high voltages to phosphor samples: the sample can be anode-grounded or cathode-grounded. The appropriate grounding depends on the purpose of the experiment. When the sample is anode-grounded, it is at the same potential as the ground and it is relatively safe to experiment near the sample. On the other hand, as all the circuits that control the electron beam are at high negative voltages, operations are generally hazardous. When the sample is cathode-grounded, the control circuits are on the ground-potential side and there is less danger to the operator. For direct measurement of the magnitude of the current of the electron beam irradiating the phosphor sample, for example by employing anode-grounding, a measuring device can be readily placed on the sample side. For modulating the beam current or applying a pulsed beam to the sample, it is advisable to use cathode-grounding.

The grounding of a vacuum envelope that contains a sample and an electron source differs, depending on which high-voltage grounding method is employed. With anode-grounding, as the sample side is earth-grounded, metal can be used for the sample holder and jigs with exception of a window observation. These metal parts and the vacuum envelope are isolated by an insulator (usually glass) from each other. As the vacuum envelope is at a negative high voltage, it is generally covered by an insulating material. A glass vacuum envelope is employed in cathode-grounding as the sample is at high voltage; the inside of the envelope is coated with conductive carbon or evaporated aluminum in order not to disturb the electron trajectories. At the same time, the outside wall of the glass envelope is coated in a similar way and is earth-grounded.

When electrons are accelerated by more than several kiloelectron-Volts, an electron gun is used as an electron-beam source. It is convenient to use a commercially available monochrome electron gun for this purpose. An oxide-coated cathode in an electron gun can generate sufficient current and the gun can modulate the current quite readily. A mixture of alkaline-earth metal (Ca, Sr, Ba) carbonates is initially coated on the cathode metal cap. By heating the carbonates under vacuum pumping, they become oxides. These oxides are activated by further heating to temperatures higher than the operating temperature. The activation process is further carried out by applying high voltage to the anode and by adjusting current flowing through the filament in a preprogrammed way while the entire system is kept under vacuum.

To evaluate a phosphor under electron-beam excitation, the beam is usually scanned over the sample. If the beam stays at the same position, the sample either becomes electrically charged or is damaged by the heat generated by electron impact. In order to avoid these undesirable effects and to simulate the conditions in a cathode-ray tube, beam scanning is recommended. The scanning is conveniently achieved using a deflection yoke placed near the top of the electron gun. In some instances, a pair of deflection plates is provided, with electron guns allowing the beam to be scanned electrostatically, but deflection sensitivity is usually low. If beam deflection or scanning is not practical in a given measurement system, the sample can be excited using DC with very low current or using a pulsed beam to achieve excitation conditions similar to those of beam scanning.

The basic principle of measurement of a phosphor by low-energy electron-beam excitation is similar to the high-energy system described above. Since measurements are done at a low voltage and the beam is irradiated evenly over the surface of the sample, the procedures are much simpler. As an example, the structure of a test tube having the same configuration as that of a triode tube is shown in Figure 22. The filament is oxide-coated, as in small vacuum tubes, and the magnitude of the current irradiating the anode sample is controlled by the grid.

Chapter one: Measurements of luminescence properties of phosphors

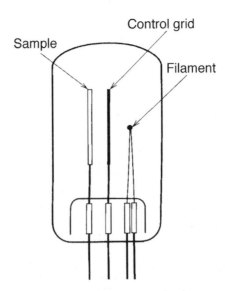

Figure 22   Structure of a low-voltage electron beam excitation test tube.

Since electron-beam excitation is carried out in vacuum, the degree of vacuum greatly affects the measured characteristics of the sample and the cathode. The vacuum must always be maintained higher than 10^{-4} Pa. When it is necessary to break the vacuum, the cathode and the heating filament should be cooled to room temperature. Care must be taken to keep the electron gun clean (i.e., free from dust in the air and other contaminants) in order to avoid high-voltage discharges.

1.1.4   Some practical suggestions on luminescence measurements

1. The environment surrounding a monochromator must be kept at a constant temperature and as low a humidity as possible. Attention must be paid to keeping mechanical shocks from the monochromator.
2. All optical paths must be vibration- and shock-free.
3. The spectral sensitivity of the photodetector must reflect the spectral range of interest.
4. A diffraction-grating monochromator is sensitive to the direction of light polarization, so light from the sample should be either unpolarized or polarized to match the polarization due to the grating.
5. A light source, photodetector, and amplifier subject to temperature drift must be warmed to attain stability before making measurements.
6. By selecting a proper filter, higher-order spectral components from the excitation source and from the monochromator should be eliminated.
7. The system must be light tight to stray light.

1.2   Reflection and absorption spectra

1.2.1   Principles of measurement

Reflection and absorption spectra measure the wavelength dependence of the intensity of light absorbed near the sample surface and in the bulk of the sample, respectively. By measuring the reflection and the absorption spectra, the absorbance of light energy by the

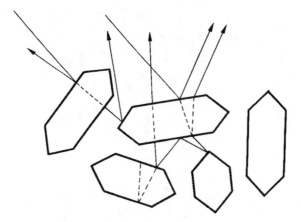

Figure 23 Light scattering by powder layers.

sample can be obtained. From the absorbance data, the energy bands of the material and the impurity levels within the material can be determined. The luminescence spectrum reveals only energy levels related to light emission. The absorption spectrum, on the other hand, gives energy levels that may or may not be involved in light-emitting transitions. In a practical sense, light absorption by a phosphor is important because the phosphor's body color greatly influences the picture contrast in a color cathode-ray tube.

When the absorption and reflection spectra of a powder phosphor sample—as opposed to a transparent solution or solid—are measured, a special experimental technique must be applied to collect the light since the powders scatter the light. A detailed theoretical discussion of the optical properties of a powder sample is given in 3.

In this section, methods to measure the absorption and reflection spectra of a sample that does not luminesce are described. There are three components in the reflected and transmitted light from a powder sample: (1) light that is deflected or scattered after being partially absorbed by the sample; (2) light that is totally reflected by the surface of the sample (specular reflection); and (3) light that passes through the gaps of the sample (see Figure 23). The first light component is what must be measured. As this light component is diffuse due to scattering, the detector only detects light within the spatial angle subtended by its aperture, and its signal decreases as the distance between the detector and the sample is increased. The intensity of the unscattered light, on the other hand, is independent of this distance if the light is collimated.

The reflectance R and/or transmittance T of a powder sample can be expressed as:

$$R \text{ or } T = \frac{I}{I_0} \tag{13}$$

where I is the intensity of the reflected or transmitted light, while I_0 is the intensity of the light source. I is the sum of intensities of the direct light I_s and the scattered light I_d, or

$$I = I_d + \alpha I_s \tag{14}$$

where α is the damping factor of the scattered light whose intensity is reduced. For the reason discussed above, generally $\alpha \ll 1$. In order to obtain good experimental results, it is desirable to have α as close to unity as possible and/or to minimize I_s. An experimental system with α is close to unity can be achieved when I_d and I_s are scattered to the same degree. Alternatively, the ratio of the intensity of scattered light between the sample and a standard material that does not absorb light at the wavelength of interest is:

$$R \text{ or } I = \frac{\alpha(I_d + I_s)}{\alpha I_0}$$

$$= \frac{I_d + I_s}{I_0} \quad (\alpha\text{: damping coefficient})$$

(15)

The scattered light includes both the specularly reflected and absorbed light components. The specular reflected light can be described by Snell's law. If the complex reflective index and Fresnel's formula are applied to Snell's law, the reflection index approaches unity as the absorption coefficient and reflective index increase. The larger the reflection index, the larger the specular component becomes and the less light absorbed by the sample. As the average particle diameter of the powder sample becomes smaller, the number of reflective surfaces increases, resulting in less penetration of light into the sample layer. This is why the body color of a sample fades when the sample is ground down to smaller particle size. When the sizes of the powder particles become as small as the wavelength of light, the intensity of scattered light has a wavelength dependence like that observed in Rayleigh scattering by gaseous molecules.

The case described above applies to a sample that does not luminesce or only luminesces weakly under excitation. The following is the experimental procedure when luminescent light from the sample is significant.

When the excitation wavelength region and luminescence wavelength region are well separated, the scattered light is observed through a filter that absorbs light in the excitation wavelength region. When the wavelength regions of excitation and luminescence overlap, special techniques must be employed. Generally, excitation light absorbed by a sample is more energetic than the emitted light. This decrease in light energy is called the Stokes' shift[20] and must be considered in most measurements. Light from the excitation source is passed through a monochromator to reduce it to a sufficiently narrow bandwidth. The scattered light is analyzed through another monochromator set at a wavelength similar to that of the first monochromator. The bandwidth of the latter monochromator should be narrow enough as to be within the range of the Stokes' shift. Thus, absorption and reflection spectra without the luminescent light component are obtained.

1.2.2 Measurement apparatus

A spectrophotometer is employed to obtain the absorption and reflection spectra of samples. The configuration of the spectrophotometer is shown in Figure 24. The spectrophotometer shown is a double-beam type and the intensity of the monochromatized sample beam is compared to a reference beam as the wavelength is scanned. An automatic spectrophotometer generates a spectrum on a recorder or has the ability to output spectral data after appropriate computation.

Generally speaking, a DC measurement of the photocurrent is susceptible to the influence of stray light and to the drift of electronic circuits. It is therefore advantageous to use an AC measurement method, as described in 1.1.2.3 utilizing a lock-in amplifier. In the AC method, the light beam is passed through a light chopper; the chopped beam is used alternately as the sample beam and the reference beam, respectively. The beam is

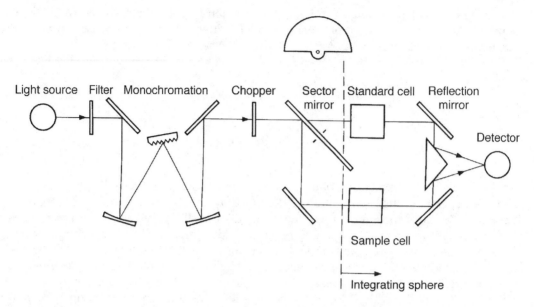

Figure 24 Optical arrangement of a double-beam spectrophotometer.

switched by a sector mirror. The two alternate beams are then detected by a photodetector. The electric signals are separated by a phase separator and then compared to each other.

There are many ways to compare the sample signal and the reference signal. When a photomultiplier is employed as the detector, comparison is achieved by adjusting the photomultiplier gain by changing the applied anode voltage. In other words, when the radiance of the reference beam at a given wavelength λ is $E_0(\lambda)$ and the detector gain is $A(\lambda)$, satisfying Eq. 16.

$$A(\lambda) \cdot E_0(\lambda) = V_c \qquad (16)$$

Since $A(\lambda) = V_c/E_0(\lambda)$ and if the sample beam $E(\lambda)$ is detected at this gain, the output signal becomes:

$$V_0 = A(\lambda) \cdot E(\lambda) = \frac{E(\lambda)}{E_0(\lambda)} V_c \qquad (17)$$

Consequently, the ratio of the sample beam to the reference beam is obtained. The above method of obtaining the ratio of the sample beam to the reference beam is useful in the wavelength range where photomultiplier tubes can be employed. For photodetectors of the thermoelectric type (thermopile) or photoconductive type (PbS and others) used in the infrared to far-infrared region, this method of comparison is difficult to apply.

As an alternate way to controlling the detector gain electrically, an optical wedge or an optical comb can be used to reduce the intensity of the reference beam. The system is designed so that the magnitude of the optical-wedge movement is proportional to the amount of light intensity reduction. When the optical wedge moves to make the intensity of the reference beam equal to that of the sample beam, the magnitude of the movement represents the light absorbance. An absorption spectrum is obtained by plotting the optical-wedge movement in the wavelength being scanned on a strip-chart recorder.

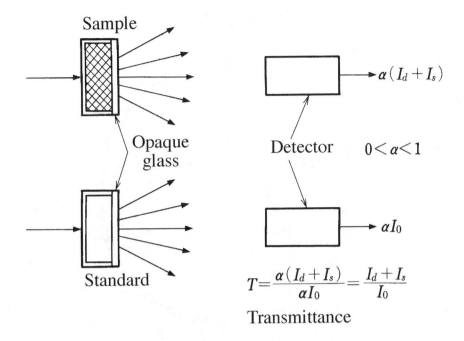

Figure 25 Absorption spectrum measurement of a powder sample by means of sample holder with opaque glass.

When the absorption spectrum of a powder sample is measured, it is important to correct for the effects of light scattering by the sample. As is expressed in Eq. 15, in order to make the direct light I_d scattered equal to the scattered light I_s, an opaque glass is employed, as shown in Figure 25.[21] When the absorption spectrum is measured, light exiting through the sample is completely scattered by an opaque glass plate placed on the sample holder. With this arrangement and since the light I_d has already been scattered by the sample layer, the degree of scattering is not changed significantly by the opaque glass plate. By comparing I_s and the light directly from the source I_0 scattered by the opaque glass, the condition required for Eq. 15 is satisfied. This technique makes the signal intensity low, so special attention must be paid to increasing the signal-to-noise ratio. As the opaque glass absorbs the infrared and the ultraviolet, this technique is not applicable to measurement in these regions.

An integrating sphere shown in Figure 26 is often employed for high-precision measurement of absorption and reflection spectra of powder samples.[22] The entire inner wall of the integrating sphere is coated with MgO or $BaSO_4$ powder. These powders have highly uniform reflectivity over a wide wavelength region, so incident light is evenly diffused. If the area of the entrance window is smaller than that of the inner wall, the light intensity anywhere on the wall can be regarded as constant.

When an integrating sphere is employed for absorption or reflection spectral measurements, the sphere is located in front of a spectrophotometer and the sample is attached to the interior wall with the excitation light shining directly on it. If a double-beam measurement is carried out, a standard white reflectance plate is provided on the sphere wall, together with an entrance window for the reference light beam. Although the inner wall of the sphere can be regarded as a perfectly diffusing surface, the resultant light may still contain a minute contribution from the direct excitation and reference beams. In order to reduce these stray components, the detector is placed perpendicular to the line between the window and the sample. To further reduce the direct beam

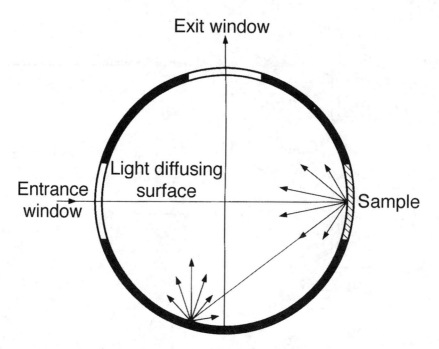

Figure 26 Integrating sphere.

contributions, the place where the direct reflecting beam shines is painted black or a black hole is positioned there.

1.3 Transient characteristics of luminescence

1.3.1 Principles of measurement

Transient properties of the luminescence of a phosphor are as important as the spectra in elucidating luminescence mechanisms. By measuring transient properties, information about the lifetimes of luminescence levels and the effective nonradiative relaxation processes can be obtained. For practical applications, these measurements yield the decay time of luminescence of the phosphor and changes in emission color with time. To obtain information about the transient properties, two types of measurements should be conducted: fluorescence lifetime and time-resolved spectroscopy.

Assuming the radiative transition rate from a fluorescent level to the ground state is R_r, and the nonradiative transition rate between the same levels is R_n, the fluorescence intensity at time t, $I(t)$, is given by:

$$\frac{d(I)}{dt} = -(R_r + R_n) \cdot I(t) \tag{18}$$

Assuming the initial condition to be $I(0) = I_0$, the above equation becomes:

$$I(t) = I_0 \exp\left[-(R_r + R_n) \cdot t\right] \tag{19}$$

The average lifetime τ measurable at this energy level can then be expressed as:

$$\tau = \frac{\int_0^\infty t \cdot I(t)dt}{\int_0^\infty I(t)dt} \tag{20}$$

By substituting the above equation into Eq. 19, one obtains:

$$\tau = \frac{1}{R_r + R_n} \tag{21}$$

From the above equation, in the ideal case where no nonradiative relaxation exists, $\tau_r = 1/R_r > 1/(R_r + R_n)$, then the experimental τ value becomes shorter than τ_r. The luminescence intensity can then be written, taking into account the time dependence of the excitation energy change $E(t)$, as follows:

$$I(t) = \eta \int_0^\infty E(t - t') \cdot D(t')dt' \tag{22}$$

where η is a constant and $D(t)$ is the decay function. For experimental purposes, the condition $E(t) = \delta(t)$ should be adhered to as close as possible, but the resulting waveform may be affected if the sample has a very short lifetime.

Measurement of a time-resolved spectrum is similar to conventional spectrum measurements. However, the time-resolved spectrum is observed instantaneously at a given time right after the excitation source is turned off. The relation between instantaneous luminescence intensity $I(\lambda,t)$ at given time t and wavelength λ can be written as:

$$I(\lambda) = \int_0^\infty I(\lambda, t)dt \tag{23}$$

1.3.2 Experimental apparatus

All the components in a system for measuring transient luminescence phenomena—that is, the light detector, the amplifier, and the analyzer—must have fast response times. When high-speed phenomena of less than a few microseconds are being measured, the response time of the entire circuit becomes critical. The signal-transmission impedance, the signal-line impedance, and the terminating impedance in the circuit must match or the signal waveform will be distorted. The pulse duration of the excitation source must also be very short.

1.3.2.1 Detector

A typical photomultiplier tube with high gain and fast response time (a rise time of a few nanoseconds) is suitable for experimental purposes. When a photodetector such as a photomultiplier tube is connected to an oscilloscope or to an amplifier, the tube's load resistance must be 50 Ω, which is the impedance of most high-frequency measuring devices. On the other hand, the maximum current that can be drawn from a photomultiplier tube is typically of the order of several hundred microamperes, so that the output voltage generated in the 50-Ω load resistance is only a few millivolts.

A convenient method of obtaining a larger output voltage from a photomultiplier tube is to connect a high-speed IC amplifier as close as possible to the tube, converting the output impedance of the tube to 50 Ω. With this technique, it is difficult to reduce the stray capacity between the photomultiplier tube anode and the amplifier to a level below several pF. The response time of the detection circuit is of the order of a few nanoseconds when the input impedance of the IC is set to exceed 1 kΩ. This is the reason a photomultiplier tube cannot monitor changes in signal waveform that are faster than a few nanoseconds, even if a high-speed IC amplifier or an oscilloscope is employed. Another disadvantage of using a photomultiplier tube to observe high-speed phenomena is that around 10 nanoseconds of delay time is required for electrons to travel between the dynodes.

As is stated in 1.1.2.2, the maximum current that can be drawn from the photomultiplier tube in the region of linear response (the region where the photocurrent is proportional to incident light intensity) is several 100 μA for usual DC light measurements. For observation of high-speed light pulses, care still must be taken not to exceed these photocurrent limits. If the photocurrent exceeds the maximum current value, there will not only be a nonlinear relationship between the light intensity and the photocurrent, but also random multiple output pulses will be generated by a single light pulse. Generally, side-on type photomultiplier tubes draw more current than head-on types.

For use when subnanosecond response time is required, a photomultiplier tube equipped with a microchannel plate and/or a biplanar phototube is available (although the biplanar phototube's sensitivity is less than that of the photomultiplier).

Linear optical detectors described in detail in the 1.1.2.2 are quite useful for measuring fast transient phenomena. The linear detector is mounted on the exit focal plane of the monochromator. The output of individual photocells are stored in the corresponding CCD memory; the data are then transferred in serial form to a personal computer or to an oscilloscope through an A/D converter.

Using a combination of a two-dimensional CCD photodetector array and an image intensifier equipped with vertical deflection electrode, a streak camera type detector can be constructed.[23,24] By choosing the proper photocathode, these devices can cover a very wide spectral range, from X-rays to the near-infrared at 1.6 μm. Streak cameras are now commercially available with time resolutions of 2 ps for synchro-scan and 200 fs for single-sweep type units. The image intensifier is particularly suited for observation of high-speed transient phenomena, as it has an inherent built-in light shutter. The sensitivity of this arrangement is comparable to that of a photomultiplier tube. A multichannel type detector in combination with a monochromator is convenient for time-resolved spectroscopy, as spectroscopic data can be obtained even with single-pulse excitation. Presently, time-resolved spectra can be obtained with ~15-ps resolution.

1.3.2.2 Signal amplification and processing

To measure an analog signal, a boxcar integrator can be used. To measure a digital signal, on the other hand, a photon counter is employed in the boxcar integrator mode. A transient recorder holds transient data after the signal is digitized and this technique can be used to record single-shot phenomena. A transient recorder stores light intensity data as a function of time, whereas a multichannel detector stores light intensity changes as a function of wavelength.

Boxcar integrator. The basic circuit of this apparatus is a sample-and-hold circuit, which is shown in Figure 27. Signals are accumulated in the capacitor C with time constant τ = RC by opening and closing switch S synchronized to a repetitive signal. The output signal V(t) can be expressed by the following equation:

$$V(t) = V_0 \left[1 - \exp\frac{-t}{\tau} \right] \tag{24}$$

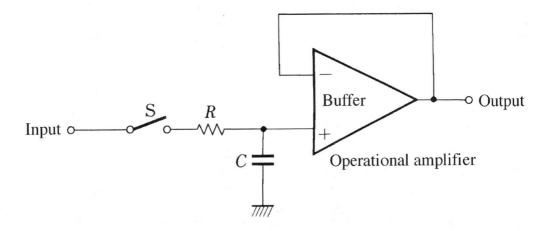

Figure 27 Sample-hold circuit diagram.

where V_0 is the final achievable value. With the time the switch is open as t_s and number of samplings per unit time as N, then $t = N \cdot t_s$. When $N \approx \tau/t_s (t = \tau)$, the output signal is about 63% of the input signal. For the output signal to be more than 99% of the input signal, the time required to accumulate the signal in the capacitor is $t \geq 5\tau$ and the number of samplings is $N \geq 5\tau/t_s$. Since the signal-to-noise ratio is defined as $\sqrt{\tau/t_s}$, the noise can be reduced by taking a smaller t_s and a longer τ. In order to obtain a good signal-to-noise ratio, therefore, the measuring time must be extended.

To measure the entire signal waveform, the timing for the opening of the gate is delayed and the delay time is stepped to cover the temporal extent of the signal. When the delay time is kept constant while the spectrometer wavelength is being scanned, a time-resolved spectrum is obtained. An electronic switching device such as an FET is often employed. Taking advantage of the fact that a photomultiplier tube is a high-gain and high-impedance current source, sampling can be done only when a high voltage is applied to the photomultiplier tube. The output current from the photomultiplier tube is then accumulated in the capacitor.

Photon-counting apparatus. When measuring light of extremely weak intensity, the analog technique described above requires long measuring times. To overcome this disadvantage, a photon-counting technique using the boxcar mode is employed. As with a boxcar integrator, intensity changes as a function of time can be obtained by varying the delay time of the photon-counting gate. The measured value can be stored in a digital memory device. Digital storage, unlike analog signals stored in boxcar integrators, is not subject to current leakage. The digital data thus obtained are convenient for further data processing.

Transient recorders. This device stores the waveform as digital data, from analog signals that are converted by a high-speed analog-to-digital converter. As this device can store data produced by a single-shot transient phenomenon, it is particularly useful for measuring chemically unstable samples. The device is normally equipped with a microprocessor that can process and manipulate the stored data; these functions include smoothing. The digital signals can be redisplayed on an oscilloscope after they have been converted back to analog by a digital-to-analog converter. Some recent commercial oscilloscopes have these capabilities built in and are convenient for optical measurements.

A transient recorder device in conjunction with a multichannel detector is useful for measuring the time-resolved spectrum of the sample, since the multichannel detector can monitor the intensity data as a function of the wavelength. A multichannel detector equipped with an image intensifier, which functions as light shutter and signal amplifier, is often employed in these cases. (See 1.1.2.)

1.3.2.3 Pulse excitation source

A conventional discharge lamp containing a rare gas such as Xe and pulsed lasers usually are employed as the excitation source. A rare-gas discharge lamp is convenient to use, as its emission light has a wide wavelength range but its power output is much lower than that of a laser. A discharge lamp with a rare gas can generate pulses of a few microseconds. When hydrogen or deuterium gas is introduced, the pulse width can be reduced to several nanoseconds. The pulse width of the discharge lamp can be further reduced to a single nanosecond by changing its electrode configuration.[25]

A pulsed laser generates a high-intensity, short pulse-width light. Detailed description of lasers are found elsewhere.[16]

For electron beam excitation, an electron beam pulse can be obtained by applying a pulsed voltage to the control grid of the electron gun. With a conventional electron gun, it is necessary to apply 100 to 200 V to the grid. Pulse generators having rise times of several tens of nanoseconds are readily available commercially. If a shorter pulse width or rise time is required, a pulse generator equipped with a thyratron is necessary.

When using the equipment described above, the following should be noted. The discharge lamp, laser, and electron beam all generate radio-frequency noise when they are activated by high-speed, high electric-power switching. Since a small signal is commonly measured by a high-impedance detector in optical studies, the system tends to pick up radio-frequency noise readily. Noise of electromagnetic origin can be picked up by signal cables, which act as an antenna. It is necessary, therefore, to have a large grounding area so that the signal cables do not form closed loops. The noise source and the detector system must also be spatially separated from each other and each individual system must be electromagnetically shielded. When a trigger signal from the excitation source to the detector system is required, either light from the excitation source is used or the trigger signal is transmitted through an optical-fiber line.

1.4 Luminescence efficiency

1.4.1 Principles of measurement

The luminescence efficiency of a phosphor concerns the radiation from the sample in the ultraviolet, visible, and infrared regions and excludes heat or X-ray radiation. Luminescence efficiency is defined as the ratio of the energy (quanta) required to excite a phosphor to the energy (quanta) emitted from the sample. Luminescence efficiency is expressed in terms of either energy efficiency (watts/watts) or quantum efficiency (photons/photons), depending on the application.

Since the radiation emitted from the phosphor sample must be measured over the entire range of its emitted wavelengths, a detector whose sensitivity is independent of wavelength should be chosen, except when only a limited wavelength range is of interest or when the luminescence efficiency between samples having the same spectral distribution is compared. A thermopile, whose spectral sensitivity is independent of the wavelength over a wide range, is normally employed to measured energy efficiency. For quantum efficiency measurements, Rhodamine B, whose excitation wavelength dependence on the quantum efficiency is known to be constant when it is excited by a shorter-

Chapter one: Measurements of luminescence properties of phosphors 35

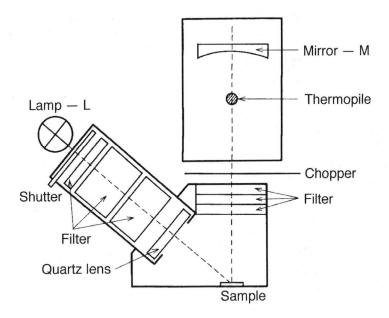

Figure 28 Luminescence efficiency measurement apparatus. (From Bril, A. and de Jager-Veenis, A.W., *J. Res. Natl. Bur. Stand., A*, 80A, 401, 1976. With permission.)

wavelength light than its emission wavelength, is often used.[26] A photon quanta-meter measures the light emitted from a phosphor sample irradiated on wavelength conversion materials such as Rhodamine B. When emitted light from a phosphor is expressed as spectral radiance distribution, the spectral distribution of photons and the total number of photons are obtained.

1.4.2 Measurement apparatus

1.4.2.1 Ultraviolet excitation

A typical apparatus for measuring the luminescence efficiency of a phosphor is shown in Figure 28.[27] The optical filters located in front of the excitation source transmit light between 250 and 270 nm, so they can isolate the 254-nm line from a high-pressure mercury lamp. A combination of a chlorine gas filter (4 atm and 4 mm thick), a nickel sulfate aqueous solution (500 g l^{-1} of NiSO$_4$·6H$_2$O in a cell 1 cm thick), and a Schott UG5 glass filter are used.[28] The output filter is a combination of glass filters that transmit the emission light but absorb the excitation light.

In the conventional method of measuring the luminescence-energy efficiency, a standard material of known energy efficiency is compared with the sample whose energy efficiency is to be measured. The energy efficiency can readily be obtained by comparing the luminescence intensity of the phosphor with that of the standard material. Standard materials are supplied by NIST (The National Institute of Standards & Technology, U.S.)[28,29]; sodium salicylate (for the far-ultraviolet and ultraviolet regions),[30] Ekta S10,[31] and Lumogen T red GC[32] are also available.

In order to measure absolute energy efficiency, the following procedure is employed using the apparatus shown in Figure 28.

Step 1. The diffuse reflectance intensity of excitation light is measured using a material such as BaSO$_4$. The wavelength dependence of the reflectance is known and is almost constant over a broad range of wavelengths.

Step 2. Without the filters, the sum of the intensities of luminescent light from the phosphor sample and the excitation light reflected by the sample is measured.

Step 3. Using filters, the intensity of luminescent light from the sample is measured.

The desired energy efficiency η_p and the reflectance of the sample r_p can be obtained from the output values of the thermopile at each step of the above measurement procedure. The reflectance of the standard material, the intensity of the excitation light, and the intensity of the reflected excitation light are denoted as R, I, and V_r, respectively. Designating the luminescence intensity as L, the thermoelectric power generated by both the luminescence from the sample and the excitation light reflected by the sample as V_p and the thermoelectric power generated by the luminescence light after passing through the filters as $V_{p,f}$, the following relations are obtained.

$$C \cdot V_r = I \cdot R$$

$$C \cdot V_p = I \cdot r_p + L \quad (25)$$

$$C \cdot V_{p,F} = \tau \cdot L$$

where C is the ratio of light energy to thermoelectric power and τ is the transmittance of the filters. From the above relations, the following equations can be derived.

$$\eta_p = \frac{L}{I(1-r_p)} = \frac{R \cdot V_{p,F}}{\tau(1-r_p)V_R}$$

$$r_p = \frac{R(V_p - V_{p,F}/\tau)}{V_R} \quad (26)$$

Thus, from the three measurement steps described above, the luminescence energy efficiency and reflectance of a phosphor sample can be obtained. From the measured energy efficiency, the quantum efficiency q_p can be expressed in terms of the luminescence intensity of a sample $p(\lambda)$ at a given excitation wavelength λ_{exc}

$$q_p = \frac{\eta_p \int \lambda p(\lambda) d\lambda}{\lambda_{exc} \int p(\lambda) d\lambda} \quad (27)$$

If the luminescence intensity distribution over the wavelength range $p(\lambda)$ is known, the quantum efficiency can be calculated using the above relation.

To obtain energy efficiency even more precisely, the amount of luminescence light absorbed by the sample itself must be taken into account. Suppose the reflectance of a sufficiently thick sample is R_∞; then the true energy efficiency η_i becomes:

$$\eta_i = \frac{2\eta_p}{1+R_\infty} \quad (28)$$

Chapter one: Measurements of luminescence properties of phosphors

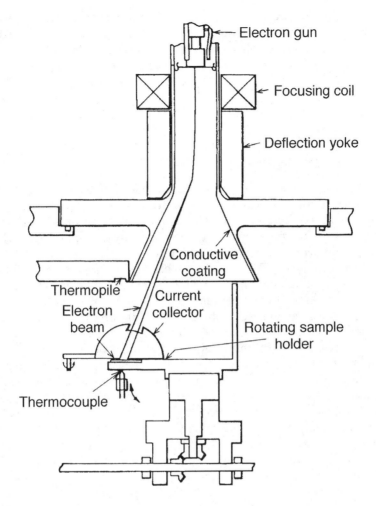

Figure 29 Luminescence efficiency measurement by electron-beam excitation. (From Meyer, V.D., *J. Electrochem. Soc.*, 119, 920, 1972. With permission.)

In the above expression, it is assumed that half of the luminescence is emitted from the phosphor surface, the other half goes into the phosphor layers, and a portion of this is reflected out in turn. The excitation light is assumed not to penetrate inside the phosphor.

1.4.2.2 Electron-beam excitation[33]

An apparatus measuring the energy efficiency of a phosphor excited by an electron beam is shown in Figure 29.[33] To measure excitation energy, one must determine not only the beam current flowing into the sample, but also the secondary electrons emitted from the sample. To measure these quantities, a semispherical current collector (Faraday cage), which has a small window to allow passage of the electron beam, is employed. When energy of the emitted luminescence from the sample is measured using a thermopile, the Faraday cage is removed.

Luminescence efficiency for electron beam excitation η can be written as:

$$\eta = \frac{2\pi \cdot r^2 \cdot C \cdot V_p}{I_0 \cdot V_0 (1 + R_\infty)} \tag{29}$$

where r is the distance between the sample and the detector, C is the sensitivity of the detector (W/W·cm²), V_p is the thermoelectric power, I_0 is the incident electron beam current, V_0 is the electron acceleration voltage, and R_∞ is the reflectance of a thick sample. The electron beam is scanned over the sample, as in TV raster scanning. Eq. 29 is satisfied when the diagonal distance of the raster scanning area is less than one-fifth of r.

1.5 Data processing

1.5.1 Spectral sensitivity correction

Since optical components such as spectrometers and light detectors used to measure the optical properties of phosphors do not have uniform sensitivity over the entire spectral range, the raw data must be corrected. The correction coefficients at each wavelength are obtained by measuring the intensity of a standard lamp with known radiation power at a given wavelength. A tungsten lamp calibrated by NIST or a halogen lamp manufactured by Ushio (JPD 100V500WCS) is available for these purposes. The correction coefficients should be obtained by measuring the standard under the same operating conditions used to measure the spectrum of a phosphor sample. These factors include parameters such as voltage applied to the photomultiplier and slit width and height of the spectrometer.

A sensitivity-corrected spectrum is generated as the product of the correction coefficients and raw spectral data. Because the correction coefficients of each optical component can vary, the entire system as a combination of components must be calibrated as a whole using a standard lamp. As the spectral sensitivity of the system also changes over time, periodic calibration of the system is recommended. When a rigorous measurement is required, corrections are made by comparing the sample's light with light from the standard lamp, point for point at each wavelength.

An example of spectral correction is shown in Figure 30. The arrow in Figure 30(b) shows Wood's anomaly, which is the non-monotonic rise of the transmission of light from a grating spectrometer as function of wavelength. This phenomenon is due to irregularities in the grating pitch. When spectra (a) and (c) in the same figure are compared, the sensitivity distortion of the spectrum becomes quite obvious.

Wavelength values as read from the counter of a scanning spectrometer are not always true values. The difference between the observed wavelength values and the true values fluctuates as the spectrometer is scanned. The wavelengths, therefore, need to be adjusted with correction coefficients, particularly when high-resolution work is being conducted. Normally, a linear approximation is used for the corrections using a few spectral lines of a mercury discharge lamp as a reference. For more precise work, a polynomial approximation is made, using as many lines as possible from several gas discharge lamps. The values for the characteristic spectral lines from the discharge lamps of various elements are tabulated in the reference books.[34]

Assuming the correction coefficients thus obtained for spectral sensitivity is $f(\lambda)$ and that of for the wavelength is $C(\lambda_0)$, the true spectrum can be written as:

$$I(\lambda) = f(\lambda) \cdot I_0(\lambda)$$
$$\lambda = C(\lambda_0) \cdot \lambda_0 \lambda = C(\lambda_0) \cdot \lambda_0 \qquad (30)$$

where $I_0(\lambda)$ is measured luminescence intensity at a given wavelength and λ_0 is the value read from the spectrometer.

Chapter one: Measurements of luminescence properties of phosphors

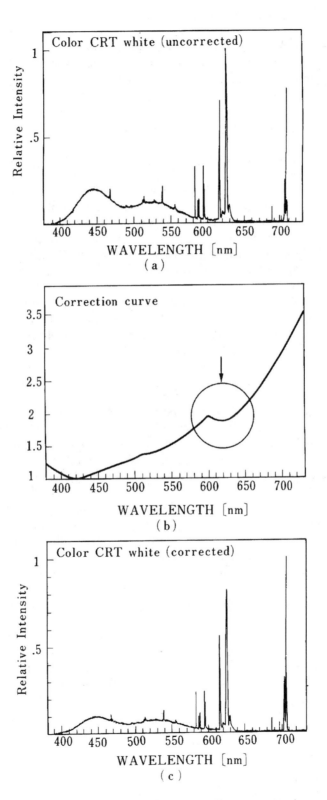

Figure 30 (a) Spectral sensitivity uncorrected spectrum. (b) Correction curve. (c) Spectral sensitivity corrected spectrum.

1.5.2 Baseline correction

Spectral data obtained from a measurement are usually superimposed on a baseline. The baseline is due to the total DC signal originating from various optical and electrical components of the system. The dark current of photomultiplier tube, stray light, and the DC bias of electronic components all contribute to the baseline value. True spectral data can be obtained by subtracting or adding the baseline to the observed spectral data. Using AC measurements described above, most of the baseline corrections can be eliminated.

The dark current value can be estimated by measuring the current at the photodetector while the optical shutter is closed. The value is recorded and used to correct the spectral data. Correction for stray light can be achieved by measuring the photocurrent for the entire wavelength region of interest without a sample and then subtracting this signal from the spectral data. This method of correcting the baseline is not applicable when the stray light intensity fluctuates. In this case, an improved measuring environment is recommended. Stray light generated within a spectrometer due to multiple reflections within the spectrometer is difficult to measure quantitatively. A spectrometer with low stray light should be selected for use.

The conventional method of correcting the baseline is to subtract the lowest signal data from the complete data. Care must be taken, however, to find the true minimum by observing the entire spectral distribution. Otherwise, small data peaks could be erroneously eliminated by this correction.

1.5.3 Improvement of signal-to-noise ratio

The shape of the spectrum and/or the decay curve can be distorted if random noise is superimposed on the data. In this case, smoothing of experimental data becomes necessary. There are two ways of smoothing the data: one is to acquire data repeatedly and then obtain the arithmetic mean of the total signal (the averaging method); another technique is to estimate the value at a point by taking the average of adjacent data points (the moving average method).

When the averaging method is applied to spectral data, the data must be reproducible as a function of wavelength. Likewise, when the averaging method is applied to decay curves, the time coordinates must be perfectly synchronized. If a simple arithmetic mean is computed for the accumulated values for each data point, the standard deviation of random noise decreases in proportion to $1/\sqrt{N}$, where N is the number of scans. (See Appendix.) The signal-to-noise ratio, therefore, can be improved to a certain extent by increasing the number of measurements. The degree of improvement brought about by this averaging, however, becomes smaller as the contribution per measurement becomes smaller and the total measuring time becomes longer.

The moving average method has an advantage over the averaging method in that it has a shorter measuring time and uses a numerical filter to smooth the experimental data. A simple moving average, $\bar{x}$, for equally spaced data x_i is defined as:

$$\bar{x} = \frac{1}{N} \sum_{i=-m}^{m} x_i, \quad N = 2m+1 \tag{31}$$

where m is an integer. The standard deviation of noise in a random noise environment is therefore proportional to $1/\sqrt{N}$, as in the averaging method. If the number of data points increases while the interval between data points is kept constant, the region over which the

Table 1 Weighted Smoothing Coefficients for 2nd- and 3rd-Order Polynomial Fits

Position	Number of points			
	5	7	9	11
−5				−36
−4			−21	9
−3		−2	14	44
−2	−3	3	39	69
−1	12	6	54	84
0	17	7	59	89
1	12	6	54	84
2	−3	3	39	69
3		−2	14	44
4			−21	9
5				−36
Normalization coefficients	35	21	231	429

average is computed is broadened and, consequently, the shape of the spectrum or decay curve is distorted. On the other hand, if one simply increases the number of data points by narrowing the data interval, the low-frequency noise component is not eliminated.

For the purpose of smoothing experimental data while minimizing the distortion of spectral data or signal-shape data, the weighted moving average method is sometimes used. Using weighting coefficients w_i, the average value can be written in analogy to Eq. 31.

$$\bar{x} = \frac{\sum_{i=-m}^{m} w_i x_i}{\sum_{i=-m}^{m} w_i} \tag{32}$$

Distortion of data can be avoided if the weighting coefficients are adjusted to yield the same result as provided by a least-squares polynomial fit. The weighting coefficients obtained in the case of second- and third-order polynomial fits are tabulated in Table 1.[35]

An example of smoothing spectral data using a combination of the averaging method and the moving average method is shown in Figure 31. In this instance, to make the signal-to-noise ratio be 10/1 using the averaging method only, the number of measurements required is 100. A similar result can be obtained using a combination of the moving average method with 9 accumulations of data (1/3) and averaging 11 data points (1/3.3).

Optimization of data processing is required even for the weighted moving average method, as unnecessary increases in the number of data points results in a distortion of the spectral-shape data.

Appendix
Standard deviation of random noise with averaging

Data sampling is used to extract multiple samples from a parent population having an average value μ, and a dispersion σ^2 defined by probability theory. If the number of samples to be extracted is N and the sample varieties are $X_i (i = 1, ..., N)$, then the sample mean $\bar{X}$ can be written as:

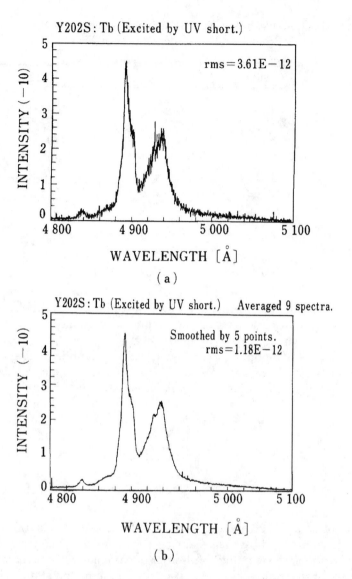

Figure 31 (a) Spectral data obtained by single measurement. (b) Smoothed spectrum after 9-times data accumulation.

$$\overline{X} = \frac{1}{N} \sum_{i=1}^{N} X_i \qquad (33)$$

In this case, the expectation value μ_N of $\overline{X}$ is given as:

$$\mu_N = E(\overline{X}) = \frac{1}{N} \sum_{i=1}^{N} E(X_i)$$
$$= \frac{1}{N} \sum_{i=1}^{N} \mu_i = \mu \qquad (34)$$

where μ_i is the expectation value of X_i. The dispersion σ_N^2 of the $\overline{X}$ is expressed as:

$$\sigma_N^2 = E\left\{\left(\overline{X} - E(\overline{X})\right)^2\right\}$$

$$= E\left\{\left[\frac{1}{N}\sum_{i=1}^{N}(X_i - \mu_i)\right]^2\right\} \qquad (35)$$

$$= \frac{1}{N^2}\left\{\sum_{i=1}^{N} E(X_i - \mu_i)^2 + \sum_{i \neq j}^{N} E\left[(X_i - \mu_i)(X_j - \mu_j)\right]\right\}$$

If X_i and X_j are mutually independent, the second term of the left-hand side of the above equation becomes 0 and the following equation is obtained.

$$\sigma_N^2 = \frac{1}{N}\sum_{i=1}^{N} E\left\{(X_i - \mu_i)^2\right\} = \frac{\sigma^2}{N} \qquad (36)$$

This means that the standard deviation σ_N of sample varieties is equal to $1/\sqrt{N}$ of the standard deviation of population σ. Consequently, the expectation value of averaging N sample data thus obtained is the true value and the component of superimposed random noise on the true value is $1/\sqrt{N}$.

References

1. Driscoll, W.G., *Handbook of Optics*, Section 1, MacGraw-Hill, New York, 1978.
2. Bass, M., Van Stryland, E.W., Williams, D.R., and Wolfe, W.L., Eds., sponsored by the Optical Society of America, *Handbook of Optics*, I & II, McGraw-Hill, New York, 1995.
3. Czerny, M. and Turner, A.F., *Z. Phys.*, 61, 792, 1930.
4. *Model 1702 Instruction Manual*, Jobin Yvon-Spex, Edison, New Jersey.
5. Seya, M., *Sci. Light*, 2, 8, 1952 Namioka, T., *Sci. Light*, 3, 15, 1954; Namioka, T., *J. Opt. Soc. Am.*, 49, 951, 1959.
6. Chemical Society of Japan, Ed., *New Experimental Chemistry Series, Fundamental Technique 3, Light [1]*, Maruzen, 1976, 165 (in Japanese).
7. *Photomultiplier Tubes Catalog*, Hamamatsu Photonics, Shizuoka, Japan, August 1995.
8. Wilson, J. and Hawkes, J.F.B., *Optoelectronics, An Introduction*, Prentice-Hall, 1989, 284.
9. *Monolithic Miniature Spectrometer, Product Information*, Carl Zeiss, Germany.
10. Wilson, J. and Hawkes, J.F.B., Reference 8, p. 296.
11. *Guide for Spectroscopy*, Jobin Yvon-Spex, Edison, New Jersey, 1994, 217.
12. Bilhorn, R.B., Sweedler, J.V., Epperson, P.M., and Denton, M.B., *Appl. Spectrosc.*, 41, 1114, 1987.
13. *Applications of Multi-channel Detectors Highlighting CCDs*, Jobin Yvon-Spex, Edison, New Jersey.
14. Nagamura, A., Mugishima, T., and Sakimukai, S., *Rev. Sci. Instrum.*, 60, 617, 1989.
15. *Discharge Lamps, Technical Brochure*, Ushio, Tokyo, Japan.
16. *CRC Handbook of Laser Science and Technology, Supplement 1: Lasers*, Weber, M.J., Ed., CRC Press, Boca Raton, FL.
17. Maeda, M. and Miyazoe, Y., *J. Appl. Phys.*, 41, 818, 1972; *Laser Handbook*, Ohm Sha, 1982 Tokyo, Japan (in Japanese).
18. Spinelli, L., Couillaud, B., Goldblatt, N., and Negus, D. K., *CLEO'91*, post deadline submission, 1991.
19. Sugahara, H., Ohnishi, Y., Matsuno, H., Igarashi, T., and Hiramoto, T., *Proc. 7th Int. Symp. the Science & Technology of Light Sources*, Kyoto, Japan, 1995.

20. Pankove, J.I., *Optical Process in Semiconductors*, Dover Publication, New York, 1975, 113.
21. *Applied Optics Handbook*, Yoshinaga, H., Ed., Asakura Shoten, Tokyo, 1973, 605 (in Japanese).
22. Chemical Society of Japan, *ibid., light[II]*, pp. 401.
23. Bradley, D.J., Liddy, B., Sibbett, W., and Sleat, W.E., *Appl. Phys. Lett.*, 20, 219, 1972.
24. Wang, X.F., Uchida, T., Coleman, D.M., and Minami, S., *Appl. Spectrosc.*, 45, 360, 1991.
25. Hundley, L., Coburn, T., Garwin, E., and Stryer, L., *Rev. Sci. Instrum.*, 38, 488, 1967.
26. Velapoldi, R.A., *Accuracy in Spectrophotometry and Luminescence Measurements*, U.S. Dept. of Commerce, Washington, D.C., 1973.
27. Bril, A. and de Jager-Veenis, A.W., *J. Res. Natl. Bur. Stand., A*, 80A, 401, 1976.
28. Bril, A. and Hoekstra, W., *Philips Res. Repts.*, 16, 356, 1961.
29. Ludwig, G.W. and Kingsley, J.D., *J. Electrochem. Soc.*, 117, 348 & 353, 1970.
30. Samson, J.A.R., *Techniques of Vacuum Ultraviolet Spectroscopy*, John Wiley & Sons, New York, 1967.
31. Grum, F., *C. I. E. Report of Subcommittee on Luminescence*, 18th Session, London, 1975.
32. Bril, A. and de Jager-Veenis, A.W., *J. Electrochem. Soc.*, 123, 396, 1976.
33. Meyer, V.D., *J. Electrochem. Soc.*, 119, 920, 1972.
34. Harrison, G.R., *MIT Wavelength Table*, The MIT Press, 1969.
35. Savitzky, A. and Golay, M.J.E., *Anal. Chem.*, 36, 1627, 1964.

chapter one — section six

Measurement of luminescence properties of phosphors

Shinkichi Tanimizu

1.6   Measurements in the vacuum-ultraviolet region 45
       1.6.1   Light sources 46
       1.6.2   Monochromators 47
       1.6.3   Sample chambers 48
       1.6.4   Measurements of excitation spectra 49
References 50

1.6 *Measurements in the vacuum-ultraviolet region*

The wavelength region between about 0.2 and 200 nm is called the vacuum-ultraviolet (abbreviated to VUV) region; most of the VUV spectrometers need to be evacuated in this region because of the opacity of oxygen in air to this radiation. Following Samson's definition,[1] the region between 100 and 200 nm is called the Schumann UV region; here, the H_2 discharge lamp can provide useful radiation as an excitation light source. The wavelength region between 100 and 0.2 nm is known as the extreme UV (EUV) region, and it includes the region of 0.2 to 30 nm, called the soft X-ray region.

    The absorption spectra of O_2 at a pressure of 10^4 Pa in the Schumann UV region, known as the Schumann-Runge bands and continuum, are shown in Figure 32.[2,3] This figure shows that the absorption coefficients of O_2 at 121.6 nm (the position of the Lyman α emission line of hydrogen atoms), and also at 184.9 nm (one of the resonance emission lines of mercury atoms) are 1 cm^{-1} or less at this O_2 pressure. These two emission lines can be used as light sources by merely flowing transparent N_2 gas along the optical path instead of evacuating the spectrometer.

    The above-referenced book by Samson (1967)[1], despite its age, is still an excellent textbook for beginners of spectroscopy in the VUV region. The book describes details of concave gratings, their mountings, light sources, window materials, detectors, polarizers, and absolute intensity measurements in the VUV region.

    Spectroscopic measurements of powder phosphors can be carried out conveniently in the Schumann UV region using LiF crystal windows, which have the shortest wavelength transmittance limit (105 nm) among any known windows. Hydrogen discharge lamps can be used as excitation light sources.

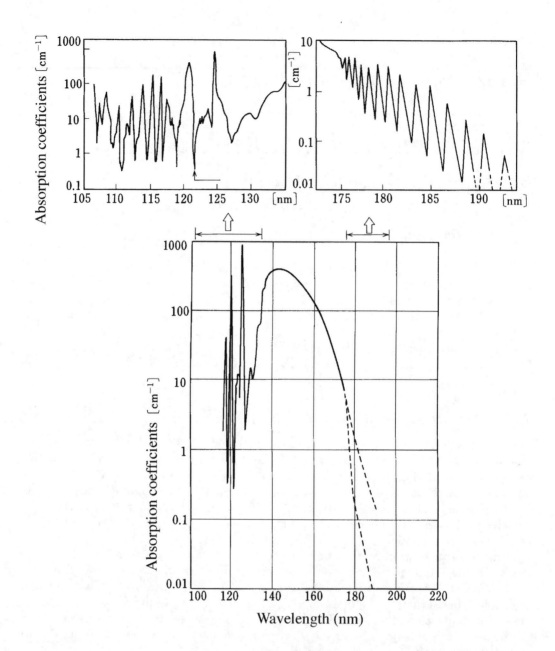

Figure 32 Absorption spectra of O_2 at a pressure of 10^4 Pa in the Schumann UV region. (From Watanabe, K., Inn, E.C.Y., and Zelikoff, M., *J. Chem. Phys.*, 21, 1026, 1953; Tanaka, Y., Inn, E.C.Y., and Watanabe, K., *J. Chem. Phys.*, 21, 1651, 1953. With permission.)

Some spectroscopic instruments and their applications in the Schumann UV region will be described.

1.6.1 Light sources

Conventional hydrogen or deuterium discharge lamps of 30 to 150 W[4] are normally used in conjunction with 0.2 to 0.4 m (focal length) VUV monochromators. Figure 33 shows the spectral output of a 30-W D_2 lamp with a MgF_2 window.[5] Emission between 165 and 370 nm

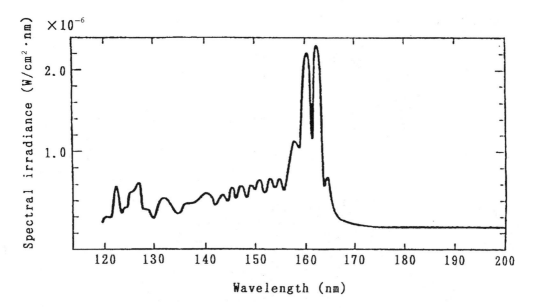

Figure 33 Spectral output of a deuterium lamp with a MgF$_2$ window in the Schumann UV region. (From Oyama, K.-I., Suzuki, K., Kawashima, M., Zalpuri, K.S., Teii, S., and Nakamura, Y., *Rev. Sci. Instrum.*, 62, 1721, 1991. With permission.)

is continuous, while below 165 nm, molecular lines predominate. The transmittance limit of polished MgF$_2$ lies at 115 nm. In these low-power lamps, the cathode (filament) and anode structures are supported in a quartz tube with dimensions of 30 to 50 mm in diameter and 70 to 190 mm in length. Starting voltages are 350 V for 30 W and 500 V for 150 W lamps, respectively. Operating currents and voltages are 0.3 A at 80 V for a 30-W lamp and 1.2 A at 120 V for a water-cooled 150-W lamp. For hot cathode lamps, the life is determined by the loss of H$_2$ or D$_2$ gas caused by a reaction with the qualtz wall. This loss is lower for D$_2$ than for H$_2$; thus, the D$_2$ lamp is widely used as an excitation light source. However, the continuous gas diffusion of D$_2$ or H$_2$ through the quartz envelopes[6] is the primary cause of the decrease in light output of the hot cathode lamps. It should be noted that the output of the D$_2$ lamp is lower than that of H$_2$ lamps below 170 nm.[7]

For 0.5 to 1 m VUV monochromators, the use of much stronger gas flow type H$_2$ lamp of 0.3 to 1 KW is recommended. Both McPherson Co. and Acton Research Co. have developed various types of VUV light sources of this type.[8] Typical parameters for cold cathode discharge lamps are: starting voltage 1.5–2 KV; operating voltage 0.5–0.8 KV, discharging current ~0.5 A, and H$_2$ gas pressure 150–300 Pa.

1.6.2 Monochromators

Seya-Namioka mount monochromators[1,9–11] are recommended for laboratory use because of the simplicity of the focusing mechanism and the possibility of using a sine drive to obtain a linear wavelength scale. As shown in Figure 34,[12] the entrance slit S_1, the exit slit S_2, and the concave grating G are placed on a Rowland circle. The grating G rotates around the vertical axis fixed at the center of G. The angle $\angle S_1 G S_2$ is set at 70°15' for an equally spaced concave grating; this configuration has the advantage of having enough space to allow the excitation light source to be set in front of S_1 and the sample chamber to be placed at the rear of S_2. The astigmatic and comatic errors of this mounting arising from large incidence angles can be greatly reduced by using a mechanically ruled aberration-corrected concave

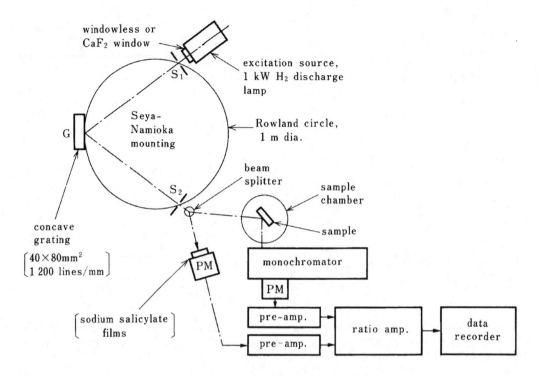

Figure 34 Schematic diagram of equipment used to measure excitation spectra of powder phosphor samples. (From Tanimizu, S., unpublished results. With permission.)

grating[13] or a specially designed holographic grating. Holographic gratings are interference gratings recorded with the use of wavefronts; in particular, the use of aspheric wavefronts reduces coma aberration in the Seya-Namioka mounting.[14]

Apart from the Seya-Namioka mount, the Johnson-Onaka mount,[15,16] which is of the normal incidence type, has been widely used for measurements of optical constants, n and k, in semiconductors.[17] High resolution can be achieved by this mount. However, it is necessary to design both the excitation source and sample chamber as compact as possible, inasmuch as they are located on the same side of the monochromator tank.

1.6.3 Sample chambers

The monochromatic light beam is divided into two beams at the rear of S_2 by a beam splitter consisting of a reflecting toroidal mirror pile; one beam excites the sodium salycilate film and the other enters into a sample chamber. Sodium salycilate is used to convert VUV light to visible light. It emits blue luminescence peaking at 420 nm with about a 10-ns decay time, and has a nearly constant quantum efficiencies of ~60% for wavelengths between 90 and 350 nm.[1,18–20] Its luminescence peak coincides with the maximum sensitivity of common photomultipliers so that the VUV radiation from the H_2 discharge lamp can be detected efficiently. Sodium salycilate films can be easily prepared by spraying a saturated methyl alcohol solution of sodium salicylate on a glass substrate kept at a temperature of 90 to 110°C. The optimum surface density of a film is about 1 mg cm^{-2}.

The chamber (inner size; 190 mmϕ × 160 mm) is equipped with a vertically rotatable turret located near the center, a small goniometric stand for mounting a detector and filters, and vacuum-sealed windows for detecting the emission, transmission, and reflection

Chapter one: Measurement of luminescence properties of phosphors

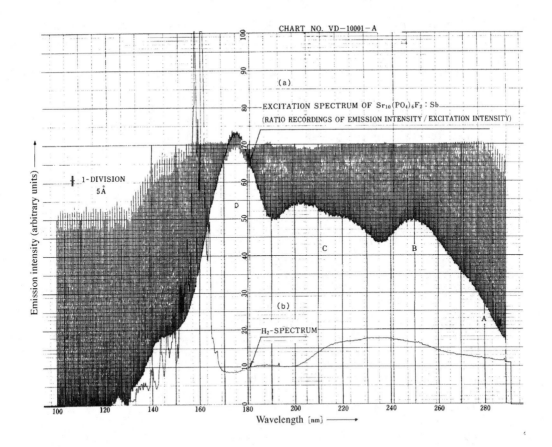

Figure 35 Recordings of an excitation spectrum for $Sr_5(PO_4)_3F:Sb^{3+}$ and the H_2 spectrum. (From Tanimizu, S., unpublished results. With permission.)

beams. A powder sample is packed into a hole of about $10 \times 10 \times 1$ mm^3 drilled at the center of a stainless steel sample holder of dimensions of the order of $25 \times 35 \times 2$ mm^3; the sample is then made flat by a glass plate. It is convenient to mount four sample holders at a time on the turret before evacuating the vacuum system. By opening or closing a gate valve located between the grating and the exit slit, one can change samples in about 10 minutes.

1.6.4 Measurements of excitation spectra

The excitation of powder phosphor samples is made at 45° incidence, and the relative output from the samples as a function of wavelength is determined in comparison with the output from the sodium salyciliate screen.

Figure 35[21] shows an example of an excitation spectrum with subnanometer resolution for the 500 nm emission of $3Sr_3(PO_4)_2 \cdot SrF_2:Sb^{3+}$ [$= Sr_5(PO_4)_3F:Sb^{3+}$]. Here, the signals from the sodium salyciliate screen and the powder sample are detected by a pair of EMI 9789QB (bialkali) photomultipliers[22] of the head-on type. The normal A, B, C, and D excitation bands of a typical s^2-type, Sb^{3+} are observed. Scanning from longer wavelengths to shorter wavelengths was found to be effective in the prevention of color center formation in the sample. In order to determine the quantum efficiency from Figure 35, it is necessary to measure a reflection spectrum of the sample in order to calculate the rate of photon absorbtion. The combination of a side-view photomultiplier coated with sodium salicylate and optical filters, both of which are mounted on the above-mentioned goniometric stand, are useful for angular-dependent reflection measurements on the sample.

In the EUV region below 100 nm, synchrotron radiation spectroscopy is very useful; this type of spectroscopy is now entering the third-generation development phase.[23] The instrumentation used in this spectroscopy (such as optical systems, detectors, electronics, and data acquisition) has been extensively reviewed. Readers are referred to References 24 through 28. Reference 27 presents quantum efficiency spectra in the 5- to 25-eV region for well-known powder phosphors excited by synchrotron radiation.

References

1. Samson, J.A.R., *Techniques of Vacuum Ultraviolet Spectroscopy*, John Wiley & Sons, New York, 1967; Samson, J.A.R.,Ederer, D.L.(Eds), *Vacuum Ultraviolet Spectroscopy, I, II*, Academic Press, San Diego, 1998.
2. Watanabe, K., Inn, E.C.Y., and Zelikoff, M., *J. Chem. Phys.*, 21, 1026, 1953.
3. Tanaka, Y., Inn, E.C.Y., and Watanabe, K., *J. Chem. Phys.*, 21, 1651, 1953.
4. See the following technical data sheets (1996). Hamamatsu Photonics K.K. (Shizuoka Pref., Japan), Models L879 & L1835. Nisseisangyo Co. (Tokyo, Japan), Model H4141SV. McPherson Co. (MA, U.S.), Model 632. Carl Zeiss Jena GmbH. (Jena, Germany), Model CLD 500.
5. Oyama, K.-I., Suzuki, K., Kawashima, M., Zalpuri, K.S., Teii, S., and Nakamura, Y., *Rev. Sci. Instrum.*, 62, 1721, 1991.
6. Lee, R.W., Frank, R.C., and Swets, D.E., *J. Chem. Phys.*, 36, 1062, 1962.
7. Levikov, S.I. and Shishatskaya, L.P., *Opt. Spectrosc.*, 11, 371, 1961.
8. See the following technical data sheets (1996). McPherson Co. (MA, U.S.), Models 630 & 631. Acton Research Co. (MA, U.S.), Model CSW-772-W.
9. Seya, M., *Science of Light (Tokyo)*, 2, 31, 1952.
10. Namioka, T., *Science of Light (Tokyo)*, 3, 15, 1954; and *J. Opt. Soc. Am.*, 49, 951, 1959.
11. Pouey, M., Principles of vacuum ultraviolet instrumental optics, in *Some Aspects of Vacuum Ultraviolet Radiation Physics*, Damany, N., Romand, J., and Vodar, B., Eds., Pergamon Press, New York, 1974, Part IV.
12. Tanimizu, S., unpublished results.
13. Harada, T. and Kita, T., *Appl. Opt.*, 19, 3987, 1980.
14. Noda, H., Harada, Y., and Koike, M., *Appl. Opt.*, 28, 4375, 1989.
15. Johnson, P.D., *Rev. Sci. Instrum.*, 28, 833, 1957.
16. Onaka, R., *Science of Light (Tokyo)*, 7, 23, 1958.
17. Philipp, H.R. and Ehrenreich, H., *Phys. Rev.*, 129, 1550, 1963.
18. Allison, R., Burns, J., and Tuzzplino, A.J., *J. Opt. Soc. Am.*, 54, 747, 1964.
19. Nygaard, K.J., *Br. J. Appl. Phys.*, 15, 597, 1964.
20. Seedorf, R., Eicheler, H.J., and Kock, H., *Appl. Opt.*, 24, 1335, 1985.
21. Tanimizu, S., unpublished results.
22. See the technical data sheets of Thorn EMI Electron Tube Ltd. (Middlesex, U.K.), 1993.
23. Ishii, T., *J. Synchrotron Rad.*, Inaugural Issue, 1, Part 1, 95, 1994.
24. Koch, E., Haensel, R., and Kunz, C., Eds., *Vacuum Ultraviolet Radiation Physics*, Pergamon-Vieweig, Braunschuweig, 1974.
25. Kunz, C., Ed., *Synchrotron Radiation—Techniques and Application*, Springer-Verlag, 1979.
26. Hafmann, H., *The Physics of Synchrotron Radiation*, Cambridge University Press, UK, 2004.
27. Berkowitz, J.K. and Olsen, J.A., *J. Luminesc.*, 50, 111, 1991.
28. J-stel., T., Krupa, J-C., Wiechert, D. U., *J. Luminesc.*, 93, 179, 2001.

chapter two — sections one–three

Measurements of powder characteristics

Sohachiro Hayakawa

Contents

2.1 Particle size and its measurement 52
    2.1.1 Shape and size of particles 52
        2.1.1.1 Circularity and sphericity 53
        2.1.1.2 Shape factor and effective and equivalent diameters 53
    2.1.2 Particle size and distribution 54
        2.1.2.1 Types of particle size distribution 54
        2.1.2.2 Mean diameter of particles 54
        2.1.2.3 Particle-size distribution function 54
    2.1.3 Classification and selection of the method for measuring particle size 57
2.2 Methods for measuring particle size 57
    2.2.1 Image analysis of particles 57
        2.2.1.1 Preparation of microsection 57
        2.2.1.2 Image analysis 57
    2.2.2 Volume analysis of particles 60
        2.2.2.1 Sieving 60
        2.2.2.2 Coulter counter 60
    2.2.3 Analysis of particle motion 62
        2.2.3.1 Sedimentation method 62
        2.2.3.2 Centrifugal sedimentation method 65
        2.2.3.3 Inertia force method 65
        2.2.3.4 Laser doppler method 66
    2.2.4 Analysis of the surface area of particles 67
        2.2.4.1 Adsorption method 67
        2.2.4.2 Transmission method 69
    2.2.5 Scattering of electromagnetic waves caused by particles 70
        2.2.5.1 Light scattering method 70
        2.2.5.2 Diffraction method 72
        2.2.5.3 X-ray diffraction and X-ray scattering 72
2.3 Measurements of packing and flow 74
    2.3.1 Definition of packing 74

2.3.2 Measurements of apparent density 75
2.3.3 Measurements of fluidity 76
      2.3.3.1 Rest angle 76
      2.3.3.2 Motion angle 76
      2.3.3.3 Powder orifice 76
References 77

Powder characteristics of powders depend on their state, whether dry or wet, compressed in mold, sintered, or dispersed as slurry. Devices and methods for measurement also depend on the state of the powders. The physical properties are determined by some basic characteristics of the powders, as follows:

1. Size and shape of powder particles.
2. The packing and flow properties are partially dependent on the particle size and shape, whereas the aggregation and flow properties depend on their kinematic or dynamic behavior. These properties are called powder characteristics in the narrow sense of the word.
3. Electrical, magnetic, optical, and acoustical properties of the powder. These characteristics are determined by the intrinsic conductivity, the light scattering behavior, the surface properties, etc. of the powder. Powder characteristics in the broad sense of the word include these properties.

Each of the physical properties of the powder can be measured experimentally using methods developed for that specific purpose. As powders are found in the different states mentioned above, different techniques are employed, as will be described below.

Powder characteristics treated in this chapter are limited to topics 1 and 2 above. Optical properties are described in Chapter 3.

2.1 Particle size and its measurement

2.1.1 Shape and size of particles

If all particles are spherical or cubic in shape, one can express their size by measuring the diameter or the length of sides, but such a case is extremely rare with fluorescent powders. For irregular-shaped particles, when they are statistically almost similar in shape, one can

Table 1 Mean and Equivalent Diameter of One Particle

Term	Definition
Two axes mean diameter (m.d.)	$(l + b)/2$
Three axes m.d.	$(l + b + h)/3$
Harmonic m.d.	$3(1/l + 1/b + 1/h)^{-1}$
Enveloping rectangular equivalent diameter (e.d.)	$(bl)^{1/2}$
Square e.d.	$(f)^{1/2}$
Circle e.d.	$(f/\pi)^{1/2}$
Cuboid e.d.	$(lbh)^{1/3}$
Cylinder e.d.	$(fh)^{1/3}$
Cube e.d.	$(V)^{1/3}$
Sphere e.d.	$(6V/\pi)^{1/3}$

Note: l: length, b: breadth, h: height, f: projected area, V: volume.

compare their size by length (l), breadth (b), and height (h), or use either a mean or equivalent diameter, as shown in Table 1.

2.1.1.1 Circularity and sphericity

Generally, the shape is described by the ideal shape closest to the actual shape of the particles, i.e., sphere, cube, cylinder, needle, flake, lump, etc. It is necessary, however, to describe the shape of the particles quantitatively or with numbers, since the shape is very important in determining other physical properties. Quantities such as circularity and sphericity are therefore defined to quantify the degree of difference in shape between an ideal sphere and the actual particles, as follows:

$$\text{Circularity} = \frac{\text{Circumference of a circle whose area equals the projection area of a typical particle}}{\text{Circumference of the projection of a typical particle}}$$

and

$$\text{Sphericity} = \frac{\text{Surface area of a sphere whose volume equals the volume of a typical particle}}{\text{Surface area of a typical particle}}$$

For irregular-shaped particles for which measurements of circumference and surface area are difficult, one uses:

$$\text{Practical sphericity} = \left(\text{Volume of a typical particle}/\text{Volume of circumscribing sphere}\right)^{1/3}$$

2.1.1.2 Shape factor and effective and equivalent diameters

Various methods for measuring the particle size use the applicable law of physics, assuming that the shape of particles is spherical or of a simple shape. In this case, the shape factors are defined in terms of the relation between the representative diameter (e.g., diameter of a sphere) D_p and the particle size for the particles of interest. Generally, the mean volume and mean surface area per particle are measured, and the volume factor ϕ_v and area shape factor ϕ_s are calculated by the following equations.

$$V = \phi_v D_p^3 \tag{1}$$

and

$$S = \phi_s D_p^2 \tag{2}$$

For spherical particles, $\phi_v = \pi/6$ and $\phi_s = \pi$.

It should be noted that the numerical values of the shape factors depend on the physics laws applicable to the measurements. For example, consider the determination of the effective diameter of a powder using their sedimentation in a solution. The measured sedimentation rate is compared with the rate for ideal spherical particles of equivalent density as determined by the Stokes equation (see 2.2.3.1). The diameter of the sphere thus determined is called the Stokes diameter, and is adopted as the effective diameter of the powder particles.

Another definition of the diameter of particles that does not use the shape factor is equivalent diameter. This diameter is that of an ideally shaped particle that is comparable in size to the particle to be measured. Typical equivalent diameters are shown in Table 1.

2.1.2 Particle size and distribution

The diameter of each particle in a powder is defined in Table 1. In general, the particles forming a powder can be described in terms of a mean or average diameter. However, actual powders are made up of particles having a statistical size distribution.

2.1.2.1 Types of particle size distribution

There are two types of distribution: frequency and accumulative distributions. Besides these two, one can define the following frequency distributions for particles ranging in diameter from D_n to D_{n+1}.

1. *Number-based distribution:* the number of particles ranging from D_n to D_{n+1} within the total number of particles, Σn.
2. *Length-based distribution:* the length of the diameter of particles ranging from D_n to D_{n+1} in the total length of diameter, ΣnD.
3. *Area-base distribution:* the surface area covered by particles from D_n to D_{n+1} in the total surface area, ΣnD^2.
4. *Weight-base distribution:* the weight of the particles ranging from D_n to D_{n+1} in total weight, ΣnD^3.

Even in the same sample, the mean of the distribution (i.e., the particle size) depends on what kind of base distribution is used. Theoretically, distribution 1 above is easy to use, but distributions 3 and 4 are frequently adopted to express that characteristics of actual powders.

2.1.2.2 Mean diameter of particles

When the properties of some powders are expressed by those of a group with a diameter $\overline{D}$ representing the particle size distribution, $\overline{D}$ is called the mean diameter. Among various D values calculated from the distribution, the one most suitable for this definition is used as the mean diameter. Various expressions for mean diameters are listed in Table 2, where L is nD, S is nD^2, and W is nD^3. As an example, the length-, area-, and weight-base distributions derived on the basis of a given number-base distribution and the mean diameters of particles, D_1 to D_4, are shown in Figure 1.

For phenomena and processes that occur on the surface of particles, such as adsorption, the total surface area of particles per unit weight of powder, i.e., the specific surface area S_w, can be defined. For spherical particles in a powder with density ρ, the specific surface area S_w can be written as follows:

$$S_w = \sum\left(n\pi D^2\right) \Big/ \sum\left(n\rho\pi D^3/6\right) = (6/\rho) \sum\left(nD^2\right) \Big/ \sum\left(nD^3\right)$$
$$= 6/\rho\overline{D} \tag{3}$$

where the mean diameter $\overline{D}$ is given by the volume-area mean diameter D_3; D_3 is also called the specific surface area diameter. Eq. 3 can be rewritten as $(6/\rho)(\Sigma n/\Sigma(n/D))$, if the harmonic mean diameter D_h (see Table 2) is used instead of $\overline{D}$.

2.1.2.3 Particle-size distribution function

Attempts have been made to describe the particle-size distribution by using comparatively simple analytic functions. No set of equations can describe all the distributions. For the distribution used or encountered most frequently, some approximations can be made.

Chapter two: Measurements of powder characteristics

Table 2 Mean Particle Diameter Based on Various Standard Distributions

Mean particle diameter based on the number-base distribution			Definition formulae for mean particle diameter based on another base		
Term	Symbol	Calculation formula	Length s.d.	Area s.d.	Weight s.d.
Length m.p.d.	D_1	$\dfrac{\Sigma nD}{\Sigma n}$	$\dfrac{\Sigma L}{\Sigma(L/D)}$	$\dfrac{\Sigma(S/D)}{\Sigma(S/D^2)}$	$\dfrac{\Sigma(W/D^2)}{\Sigma(W/D^3)}$
Area-length m.p.d.	D_2	$\dfrac{\Sigma(nD^2)}{\Sigma(nD)}$	$\dfrac{\Sigma(LD)}{\Sigma L}$	$\dfrac{\Sigma S}{\Sigma(S/D)}$	$\dfrac{\Sigma(W/D)}{\Sigma(W/D^2)}$
Volume-area m.p.d.	D_3	$\dfrac{\Sigma(nD^3)}{\Sigma(nD^2)}$	$\dfrac{\Sigma(LD^2)}{\Sigma(LD)}$	$\dfrac{\Sigma(SD)}{\Sigma S}$	$\dfrac{\Sigma W}{\Sigma(W/D)}$
Weight m.p.d.	D_4	$\dfrac{\Sigma(nD^4)}{\Sigma(nD^3)}$	$\dfrac{\Sigma(LD^3)}{\Sigma(LD^2)}$	$\dfrac{\Sigma(SD^2)}{\Sigma(SD)}$	$\dfrac{\Sigma(WD)}{\Sigma W}$
Area m.p.d.	D_5	$\sqrt{\dfrac{\Sigma(nD^2)}{\Sigma n}}$	$\sqrt{\dfrac{\Sigma(LD)}{\Sigma(L/D)}}$	$\sqrt{\dfrac{\Sigma S}{\Sigma(S/D^2)}}$	$\sqrt{\dfrac{\Sigma(W/D)}{\Sigma(W/D^3)}}$
Volume m.p.d.	D_v	$\sqrt[3]{\dfrac{\Sigma(nD^3)}{\Sigma n}}$	$\sqrt[3]{\dfrac{\Sigma(LD^2)}{\Sigma(L/D)}}$	$\sqrt[3]{\dfrac{\Sigma(SD)}{\Sigma(S/D^2)}}$	$\sqrt[3]{\dfrac{\Sigma W}{\Sigma(W/D^3)}}$
Volume-length m.p.d.	D_{vL}	$\sqrt{\dfrac{\Sigma(nD^3)}{\Sigma(nD)}}$	$\sqrt{\dfrac{\Sigma(LD^2)}{\Sigma L}}$	$\sqrt{\dfrac{\Sigma(SD)}{\Sigma(S/D)}}$	$\sqrt{\dfrac{\Sigma W}{\Sigma(W/D^2)}}$
Harmonic m.p.d.	D_h	$\dfrac{\Sigma n}{\Sigma(n/D)}$	$\dfrac{\Sigma(L/D)}{\Sigma(L/D^2)}$	$\dfrac{\Sigma(S/D^2)}{\Sigma(S/D^3)}$	$\dfrac{\Sigma(W/D^3)}{\Sigma(W/D^4)}$
Geometrical m.p.d.	D_g	$\dfrac{\Sigma(n \cdot \log D)}{\Sigma n}$	$\dfrac{\Sigma\{(L/D)\log D\}}{\Sigma(L/D)}$	$\dfrac{\Sigma\{(S/D^2)\log D\}}{\Sigma(S/D^2)}$	$\dfrac{\Sigma\{(W/D^3)\log D\}}{\Sigma(W/D^3)}$

Note: m.p.d.: mean particle diameter. s.d.: standard deviation.

Normal distribution.   The relationship between the size of particles D and the integrated number n(D) from 0 to D for a normal distribution is expressed by the following equation:

$$dn(D)/dD = \left\{\sum n/\sigma(2\pi)^{1/2}\right\} \exp\left[-(D-\overline{D})^2/\sigma^2\right]$$

$$\sigma = \left\{\sum (D-\overline{D})^2/(n-1)\right\}^{1/2}$$

(4)

where Σn is the total number of particles and σ is the standard deviation of the distribution.

Logarithmic normal distribution.   This is the distribution obtained by substituting log D and log σ for D and σ in Eq. 4, respectively. In this case, the relationship between D and n(D) is expressed by:

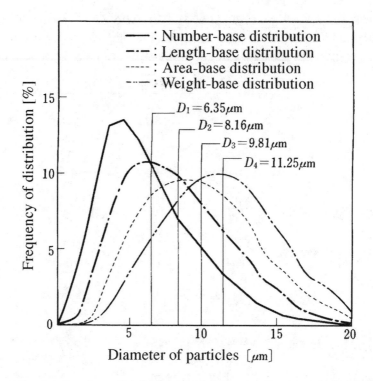

Figure 1 Example of various base distributions and the mean diameter of particles.

$$dn(D)/d\log D = \left\{\sum n / \log \sigma_g (2\pi)^{1/2}\right\} \exp\left[-\left(\log D - \log \overline{D}_g\right)^2 / 2\log^2 \sigma_g\right]$$

$$\log \overline{D}_g = \left(\sum n \log D\right) / \left(\sum n\right) \tag{5}$$

$$\log \sigma_g = \left\{\sum \left[n\left(\log D - \log \overline{D}_g\right)^2\right] / \sum n\right\}^{1/2}$$

Rosin-Ramler's equation. Rosin and Ramler proposed the following equation for the particle-size distribution of crushed coal. The percentage of particles with diameter R, larger than D, is expressed as:

$$R = 100 \exp(-bD^n) \tag{6}$$

With the substitution $b = 1/D_e^n$, this equation is rewritten as:

$$R = 100 \exp\left\{-(D/D_e)^n\right\} \tag{7}$$

n and b(or D_e) are characteristic parameters of the distribution to be determined by measurement. The parameter n is called the R-R distribution constant and has a value of 0.5 to 1.5.

2.1.3 Classification and selection of the method for measuring particle size

Measurements of the particle size commonly used at present can be classified according to the technique employed and depending on the range of particle sizes to be measured. These methods are summarized in Table 3. Various common methods for these measurements are described in the next section.

In selecting the method for measuring particle size, careful consideration must be given to the properties of the sample, the purpose of the measurement, the required measuring accuracy, etc. In general, the measuring methods are classified according to powder conditions for the ranges of particle sizes to be measured, and are shown in Table 4. In selecting the appropriate method, one needs to be aware of an approximate particle size in addition to the general physical properties of powder samples.

2.2 Methods for measuring particle size

2.2.1 Image analysis of particles

2.2.1.1 Preparation of microsection

Measurements of particle size conducted using an optical microscope require special care to insure that the particles are homogeneously dispersed in a medium and no aggregation occurs to affect the distribution in a microsection; this care is required so that the particle distribution in the sample can be considered to be representative of the whole powder. The particles may be homogeneously dispersed by sprinkling a small amount of the powder on a glass plate; but more commonly, a small amount of the powder is dissolved in a suitable dispersion medium, coated on a glass plate, and then dried. Particles can also be dispersed homogeneously by admixing the powder into a viscous resin and then thinly coating the resin on a glass plate.

For electron microscope inspections, special care is required so that the supporting lamella does not affect the distribution. As in the case above, powders may be sprinkled over the supporting film with a writing brush. Otherwise, samples dispersed in a mixture of water and linseed oil can be deposited on a supporting film; the dispersion medium is then removed by a suitable solvent, leaving a well dispersed specimen. In addition to direct observation of samples using TEM (transmission electron microscopy), observation of the morphology of particles by the shadowing method and SEM (scanning electron microscopy) observation by the metal replica method can also be used.

2.2.1.2 Image analysis

The measurement of particle size may be made directly using the visual image produced by an optical microscope; in most cases, however, sizes are measured using photographs of particles. The size of particles on the photograph is determined using an appropriate scale once the magnification factors have been established. At present, uses aided by computer systems for semiautomatic or fully automatic analysis are being developed for these measurements.

The number-base distribution, and various representative diameters, shape factors, etc. can now be determined by a semiautomatic digitizer, which measures the x and y coordinates of each particle image and analyzes these coordinates with a computer. A fully automatic image analysis apparatus are used predominantly at present. The three-dimensional analysis using shadowing and holography and the study of the morphology analysis of particles also can be conducted with these instruments.[1]

The image treatment method above can cover the broadest range of particle sizes in principle, and it is capable of determining the sizes ranging from ultrafine to coarse. The

Table 3 Classification and Characteristics of the Methods for Measuring Particle Size

Principle of measurement	Measuring method	Measurable range of size (mm / μm / nm)	Measured particle size	Distribution base	Sample condition
Image analysis	Optical microscope		Length	Number	Wet, dry
	Electron microscope		Area	Number	Dry
Volume analysis	Sieving		Sieve opening	Weight	Wet, dry
	Coulter counter		Equivalent diameter	Number	Wet
Analysis of particle motion	Gravity sedimentation		Stokes diameter	Weight	Wet, dry
(Sedimentation)	Centrifugal sedimentation			Weight	Wet, dry
	Light transmission			Area	Wet
	Air sieving			Weight	Wet, dry
	Elutriation				
(Inertia force method)	Cascade impactor		Stokes diameter	Weight	Dry
	Cyclone			Weight	Wet, dry
Surface area analysis	Kozeny-Carman method		Specific area diameter		Wet, dry
(Permeability method)	Knudsen method				Dry
(Adsorption method)	BET method		Specific area diameter		Dry
	Fluxion method				Wet, dry
	Heat of wetting				Dry
Electromagnetic wave scattering	Light diffraction		Reduced diameter of sphere	Weight	Wet, dry
	Light scattering (angle distribution)		Mean effective diameter		Wet, dry
	(Doppler width)				Wet, dry
	X-ray diffraction (Sherrer width)				Wet, dry
	X-ray small angle scatter				Wet, dry

Table 4 Powder Conditions and Methods of Measurement to be Applied

Condition	Purpose of measurement		Coarse powder (cm / mm)	Fine powder (100 μm / 10 μm / μm)	Super fine powder (100 nm / 10 nm / nm)	Diameter	Time required
Powder	Approximate	Distribution	——— Sieve ———			E	L
				——— Micros. (Opt.) ———		S	Sh
				——— Touching ———		—	Sh
				——— Bulk density ———		—	Sh
	Approximate	Mean		——— Transmission ———		E	Sh
Wet method ↓	Detail	Dis.	——— Sieve ———			E	L
				——— Air sieve ———		E	L
				——— Micros. (Opt. and Electron.) ———		S	L
				——— Transmission ———		Sp	Sh
	Detail	Mean			——— Adsorption ———	Sp	L
Suspension ↓	Approximate	Dis.	——— Sieve ———			E	L
				——— Grinding ———		—	Sh
				——— Micros. ———		S	Sh
				——— Sedi. ———		E	L
				——— Sedi. (volume) ———		—	L
	Approximate	Mean		——— Wet trans. ———		Sp	Sh
	Detail	Dis.	——— Sieve ———			E	L
				——— Micros. (Opt.) ———		S	L
				——— Sedi. ———		S	L
					——— Micros. (Electron) ———	E	L
				——— Centrifugal sedi. ———		E	L
			——— Coulter counter ———			E	Sh
				——— Light scatter. ———		E	Sh
	Detail	Mean			——— Adsorb. (Liquid) ———	Sp	L
			——— Permeability (Wet) ———			Sp	L
						Sp	L

Note: E: Effective diameter, S: Statistical diameter (particle size depends on the treatment of results obtained), Sp: Specific area diameter, L: Long, Sh: Short, Dis.: Distribution, Sedi.: Sedimentation.

relative errors of the mean particle diameter ε, however, increase with a decrease in the number of particles N being measured. In general, the relation between ε and N may be expressed by:

$$\log[\varepsilon] = -(1/2)\log N + \log K \qquad (8)$$

where K is a constant determined by the degree of dispersion, the shapes, the definition of representative diameters, and the size distribution.[2]

Recent measurements on some common samples conducted by a group sponsor by the Society of Powder Technology shows that these errors are influenced by the personal skills of the experimentalist. For a broad particle-size distribution, even when the particles are of ideal shape, more than 1000—preferably several thousand—particles need to be measured to reduce errors to a range of 2 to 3%. Highly accurate determinations require that the particle analysis be totally automated after good-quality microscope images have been obtained.

2.2.2 Volume analysis of particles

This analysis is based on the measurement of particle size using the volume of particles of some phenomenon proportional to this volume. Techniques often used are sieving and the coulter counter. Both techniques can be regarded as measurements of cross-sectional areas of particles, which can then be readily converted to a volume.

2.2.2.1 Sieving

Measurement of particle size by sieving is effective for powders composed of relatively coarse particles. In sieving, special attention should be given to the accuracy of the sizes of the sieve and also to the motion of the sieve. Methods need to be developed to estimate particle diameters, especially those of particles with shapes far from spherical. Standard sieves are produced by JIS, Tyler Co. (U.S.), ASTM, and others. JIS and Tyler Co. sieves are most widely used in Japan. Both types have sieve openings that vary in a $2^{1/4}$ geometric series. For the JIS sieves, 30 stages cover the range from 37 µm to 5.660 mm. Particles larger than 100 µm in diameter are rarely used in practical fluorescent materials. Special microsieves are manufactured for fine-grained samples down to few micrometers in size, as shown in Table 5; often, wet sieving is employed to avoid aggregation of powders.

2.2.2.2 Coulter counter

This method determines the size of particles in a suspension by measuring changes in the electrical resistance and the number of particles by counting electrical current pulses. This method is widely applied in fields such as medicine and the food and ceramics industries; for example, the size and number of blood corpuscles and bacteria are determined in this way. The measuring range of the particle sizes is between 0.25 and 500 µm.

A diagram of the apparatus used in this method is shown in Figure 2; the principle of measurement for particle sizes is described below. An electrolytic solution fills a vessel with a wall having a capillary hole that divides the solution into two parts. The electrodes are put in both sides of the divider. When a voltage is applied through the electrodes, the flow of electric current is controlled predominantly by the electrical resistance of the capillary hole. When the capillary is filled with a suspension, fine particles in the capillary region change the electrical resistance, depending on their size and resistivity. The capillary is assumed to be a cylinder with a cross-sectional area of A and a length of l. The change in electrical resistance ΔR, when a particle of volume V and cross-sectional area a enters the capillary, is given by:

Table 5 The Standard Specification of Microsieves

1. Sieve opening:												
Normal Opening (μm)		63	53	45	32	25	20	16	12.5	10	8	5
Allowance (μm)	Max.	65	55	47	34	27	22	18	13.5	11	9	6
	Min.	64	51	43	30	23	8	14	11.5	9	7	4
Porosity (%)		40	40	40	40	40	40	40	39	25	12	10

2. Material of sieves:
   Precision sieves: silver alloy plated with nickel

3. Size of screen frame:

Class	D	d	H	h
L	90	75	38	20
S	52	38	56	19

4. Material of screen frame:
   Aluminum and transparent acrylate resin

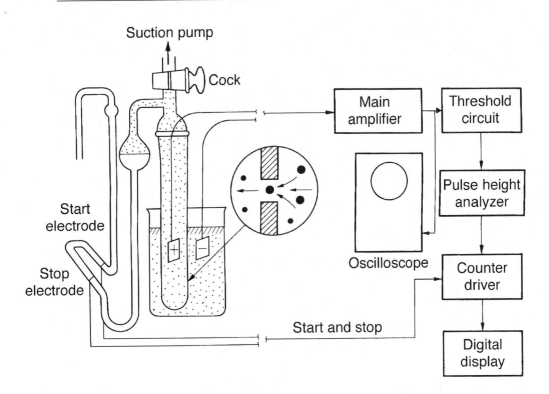

Figure 2 Coulter counter.

$$\Delta R = (\rho_0 V/A^2)\{1/(1-\rho_0/\rho) - (aD/1A)\} \qquad (9)$$

where ρ_0 and ρ are the specific resistivity of the electrolytic solution and the particle, respectively. D is the reduced diameter of the sphere approximating the irregular-shaped particle. The change ΔR is approximately proportional to V for $\rho \gg \rho_0$. The accuracy of the detector and the counter used in this apparatus for measuring the diameter of particles is ±0.01 μm.

The procedure for measurement is as follows. A particle suspension is placed in a beaker (50–400 ml in volume). The suspension is then introduced into the tube through the capillary (30–560 μm in pore diameter and 0.10–1 mm in length). Each time a particle passes through the capillary, the electrical resistance changes an amount ΔR given by Eq. 9. The pulses generated by the changes, ΔR, are amplified and the height and number of pulses are analyzed to give the particle-size distribution.[3,4]

This method is suitable for samples that are difficult to analyze using the sedimentation techniques (see 2.2.3.1); this method is suitable for samples of lower relative density or for samples containing trace amounts of powder. This method, however, is unsuitable for a sample having a fairly broad particle-size distribution, and cannot be used for powders soluble in the electrolytic solution. Particle sizes obtained in this method are not absolute, but relative; results need be compared with standard samples of known size and distribution in order to obtain absolute values.

2.2.3 Analysis of particle motion

A particle moving in a field of force such as gravity, centrifugal force, etc. experiences a retarding force when moving through a viscous medium and eventually reaches a constant velocity, i.e., the terminal velocity. Measurements of the terminal velocity provide a method for determining the size of the particle.

2.2.3.1 Sedimentation method

The sedimentation method has traditionally been used to determine the diameter by measuring the sedimentation terminal velocity v of the particles. The method for size determination depends on Stokes' law; namely, the terminal velocity of a spherical particle in a viscous liquid is given by:

$$v = (1/18)(\rho_p - \rho) g D_p^2 / \eta \tag{10}$$

where ρ_p and ρ are the density of the particle and the medium, respectively, η is the viscosity of the medium, D_p is the diameter of the particle, and g is the acceleration due to gravity.

The sedimentation rate depends upon the particle-size distribution. This distribution is determined by measuring the weight distribution as a function of depth in the vessel and the sedimentation time. The weight distribution is generally determined by the number of particles passing through a fixed depth; in some methods, such as the specific gravity balance and the light transmission methods, the variation of the distribution with depth is determined instantaneously. The method at fixed depth is classified into two types, as follows:

1. *Increment type:* the variation of particle concentration at a certain depth h (the pipette method, the light transmission method, and the specific gravity balance method) is measured.
2. *Accumulation type:* at a depth h, the variation in a quantity related to the total concentration above h or below h is determined (the hydrometer method and the sedimentation balance method).

The difference between these two techniques and an outline of analyzing procedures are shown schematically in Figure 3.

Chapter two: Measurements of powder characteristics

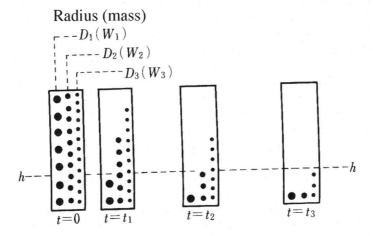

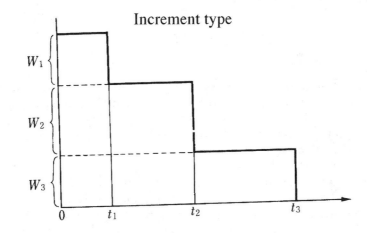

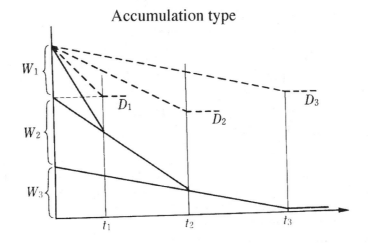

Figure 3 Analysis of the sedimentation method.

The outlines of various measuring methods are as follows:

Pipette method. A small amount of the suspension is drawn into a pipette inserted to a fixed depth h; the particle concentration of the suspension is obtained by measuring the density of the liquid. The measurement is repeated at fixed time intervals so that the variation of the particle concentration with time can be obtained. The particle diameter D_p can be derived using the Stokes equation (Eq. 10) with the substitution $v = h/t$ where t is the time when the pipette is filled. Using the particle concentration as measured above, the corresponding particle diameter gives the cumulative distribution curve. The Andreasen pipette is often used in this method because of its simplicity.

Light transmission method. This method determines the particle concentration by estimating the decay of a narrow beam of collimated light incident into the suspension. Let the intensity of incident light be I_0, the intensity of transmitted light I, and the length of the optical path in the liquid L. The logarithm of the ratio of the intensity of incident light to that of transmitted light is given by:

$$\log(I_0/I) = kcL \sum_{i}^{D_{max}} K_R n_i D_i^2 \qquad (11)$$

where c is the concentration of suspension, n_i the number of particles with the particle diameter D_i in a sample of 1 gram, and k an experimental constant that depends on the equipment and the operating conditions. The constant K_R is called the Rose's extinction coefficient due to the optical properties of particles and is usually assumed to be unity. The limit of summation is to the highest value D_{max} of D calculated by Eq. 10 using a velocity $v = h/t$ for the depth h and time t. Application of Eq. 11 is limited to the range of conditions for which geometrical optics is applicable; the minimum particle diameter measurable must be two or three times larger than the wavelength used.

While the sedimentation method provides the weight-base distribution, the light transmission method provides the area-base distribution of particles as understood from Eq. 11. Also, the relation between D_{max} and the concentration distribution can be derived by scanning the light at various depths h at a fixed time t. This method can be used to obtain the particle-size distribution very quickly.

Specific gravity balance method. If the powder suspension is allowed to settle, a concentration gradient is caused by the particle-size distribution. The particle concentration can be determined by measuring the buoyancy experience by a small sinker hung on a thin string, while changing the depth of the sinker h without disturbing the suspension. This method provides the cumulative distribution because the depth h gives the particle diameter D.

Hydrometer method. Instead of a sinker, a hydrometer (liquid densitometer) can be dipped into the suspension. The mean specific weight of the part of the suspension where the hydrometer is submerged, determines the particle concentration in that region. Changes in the concentration with time provide the particle-size distribution. This method is used to determine particle-size distribution for soils and similar substances (*JIS A 1204*).

Sedimentation balance method. This method is applicable both in the liquid and gaseous phases. The balance pan is placed at a level near the bottom of the sedimentation tube and measures the weight of all particles being deposited in time. An increase in the

Chapter two: Measurements of powder characteristics 65

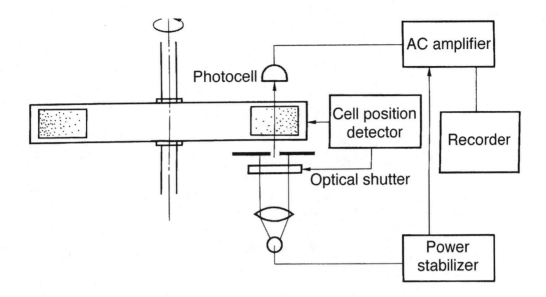

Figure 4 Equipment for measuring particle size by centrifugal sedimentation method of light transmission type.

weight of the sediments with time provides a sedimentation curve and the differentiation of the curve can provide the particle-size distribution. This method is generally used in apparatuses for particle-size measurement in which the results are automatically recorded and processed by computers.

2.2.3.2 Centrifugal sedimentation method

Fine particles that precipitate very slowly through gravitation may do so quickly under the action of a centrifugal force; as in the previous case, the terminal velocity v (=dr/dt, r is the radial position from the axis of rotation) can be written as:

$$v = (1/18)(\rho_p - \rho)\omega r D_p^2 / \eta \qquad (12)$$

where ω is the angular velocity of rotation. In this equation ωr replaces g in Stokes equation (Eq. 10).

As the relative change of the distance due to the movement of the particles is small for large r, the value of r is assumed to be a constant and results obtained can be analyzed by the same methods as in sedimentation discussed above. The suspension is disturbed, when the system starts or stops the rotation. Consequently it is desirable to measure the terminal velocity of particles while the suspension is rotating at a constant rate. Figure 4 shows the equipment used in light transmission measurements of centrifugal sedimentation. Many kinds of commercial equipment are available, and this is by far the most common method used for these measurements.

2.2.3.3 Inertia force method

The above two methods use the principle that the masses of particles control the terminal velocities in their movements. A second method uses the difference in momentum caused by the difference in the mass, although the particles might have the same velocity.

Cascade Impactor: An air current containing particles is passed through a series of nozzles; the direction of the air flow is changed by impact plates. Larger particles are

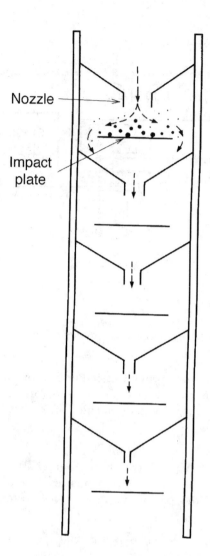

Figure 5 Cascade impactor.

collected on the plates but smaller particles are deflected sufficiently to pass through into the successive regions. By stacking the impact plates with decreasing diameters of the nozzles, as shown in Figure 5, the particles are sorted on each impactor according to their sizes. The results yield the weight-base distribution. This method is, however, purely empirical because there is no formal theory that allows the calculation of the diameter of particles; use of so-called small multistage cyclones suffers from the same problem.

2.2.3.4 Laser doppler method

The methods above analyze the movement of groups of particles from which the weight-base distribution can be deduced. The number-base distribution can also be derived by analyzing the velocity of each particle and calculating the particle mass. There is another method for counting the number of particles possessing a given sedimentation rate, i.e., using the principle of the laser velocity meter. This method can be applied to a broad range of sedimentation rates for fine particles, both in air and in fluids.

Chapter two: Measurements of powder characteristics

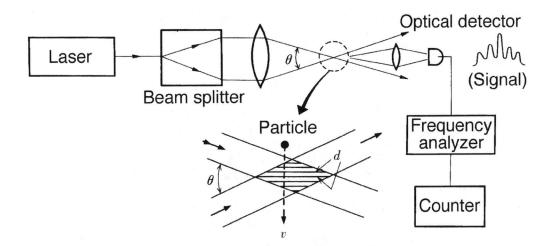

Figure 6 Laser Doppler method.

Figure 6 shows the schematic of this method. A laser beam is split into two beams that are then made to cross each other to produce an interference pattern. The width of an interference d is given by:

$$d = \lambda/2\sin(\theta/2) \qquad (13)$$

where λ is the wavelength of the laser and θ is the crossing angle. When the particles enter the interference region, they scatter light. The scattered light produces a so-called Doppler beat as shown in the right portion of Figure 6. The frequency of the beat, f_{LD}, is given by the following equation with v the velocity of particles:

$$f_{LD} = (v/d)\{2\sin(\theta/2)/\lambda\}v \qquad (14)$$

By analyzing the frequency of the beat, a signal at the frequency f_{LD} corresponds to a particle having a velocity v. This velocity v can be reduced to D_p using Eq. 10 as in the case of the sedimentation, or Eq. 12 as in the case of measurements in a centrifugal force field. A number-base distribution is obtained in this way. This method has been applied for the observation of Brownian motion of fine particles.

2.2.4 *Analysis of the surface area of particles*

The basis for measuring surface reactions and reactivity of particles is the determination of their surface area. Measurements give the surface area per weight from which the mean particle diameter can be calculated when the solid density of particles is known and the shape of particles is assumed to be spherical or granular. Common methods for these measurements are the adsorption and the wetting heat method, both of which rely on surface reaction of the solids; the transmission method is based on another principle, which will be described in 2.2.4.2.

2.2.4.1 *Adsorption method*
This method uses a solute molecular compound in liquid or a low molecular weight compound in gaseous phase as an adsorption substance. In the former, the colorimetric

method, which is simple and convenient, is used to determine the amount of dye molecules adsorbed from the dye solution onto the powders; its accuracy, however, is poor. In most cases, gaseous adsorption is used. Measurements for this method are divided into two categories: the volumetric and gravimetric methods.

Volumetric method (BET method). This method measures the amount of adsorbed substance on the basis of the change is pressure or volume due to adsorption of a gas on the particles; this process is named the BET method after its three co-founders: Brunauer, Emmett, and Teller. A known amount of powder in a sample container is heated and evacuated to remove adsorbed gases; the container is then removed from the vacuum pump system. A volume of gas, measured beforehand, is transferred into the sample container at low temperature; the amount of gas in the container can be calculated by measuring the pressure. A sample of 10 to 15 ml is generally sufficient for these measurements; this is not difficult for samples having total surface areas of more than several square meters (m²). This method, however, is unsuitable for volatile samples or for powders with low melting points, from which it is generally difficult to remove absorbed gases. The adsorption isotherm obtained by determining the relation between the equilibrium pressure and the amount of adsorption gives the surface area of the sample through the analysis described for the adsorption isotherm below.

Gravimetric method. In this method, the amount of adsorption is determined by measuring the increment in the weight of sample. For example, the increment due to the adsorption of N_2 gas molecules on a 100-m² surface is 28.6 mg. The adsorption balance method gives the highest accuracy, but the spring balance method and the cantilever method can also be used.

In order to determine the adsorption isotherm as a function of gas pressure, it is necessary to calculate the true amount of adsorption W_a using the following equation.

$$W_a = W_{as} + \rho V_s + \rho V_a - W_s \tag{15}$$

where W_{as} is the apparent weight of the powder after adsorption has taken place, W_s the weight of sample in a vacuum, V_s and V_a the volumes of sample and the absorbed gas, respectively, and ρ the density of the gas at equilibrium. For pressures not exceeding 1 atm, ρV_a is negligible.

Adsorption isotherm. In addition to N_2, Ar, H_2O, etc., various gases are used as adsorption media; N_2 is the most suitable because of its inertness and ease of use. Typical isotherms, shown as types I and II, are illustrated in Figure 7, though adsorption isotherms take various forms depending on the method of measurement. In the figure, P is the equilibrium pressure, P_0 is the saturated vapor pressure, and V is the amount of gas adsorbed. Usually, a type I isotherm indicates monolayer adsorption, whereas multilayer adsorption yields type II isotherms. Adsorption isotherms are usually reversible, but hysteresis occurs in samples with particularly strong adsorption and/or in porous samples. Equations representing adsorption isotherms for these two types, I and II, have been derived. Representative equations corresponding to type I and II in Figure 7 are as follows.

$$\text{The Langmuir equation}: \quad P/V = 1/(V_m b) + P/V_m \tag{16}$$

$$\text{The BET equation}: \quad V = V_m C_p/(P_0 - P)\{1 + (C-1)P/P_0\} \tag{17}$$

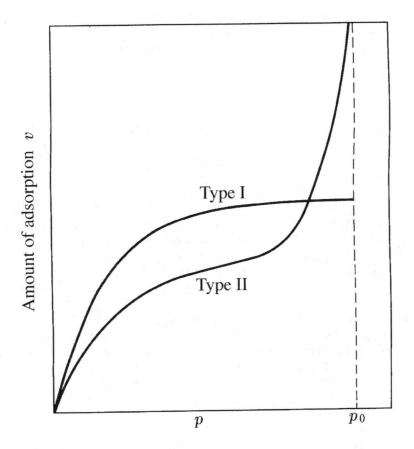

Figure 7 Types of adsorption isotherms.

where b, C, and V_m are constants determined experimentally. The constant V_m in the BET equation represents the volume of adsorption molecules necessary to form a monolayer. The BET equation can be rewritten as:

$$P/V(P_0 - P) = 1/V_m C + \{(C-1)/V_m C\}(P/P_0) \tag{18}$$

Plotting $P/V(P_0 - P)$ vs. P/P_0, the slope and the intercept give V_m if the plots are linear. The weight-specific surface area S_w (m² kg⁻¹) is given by the product of the number of gaseous molecules, calculated from V_m, and the molecular cross-section σ as:

$$S_w = V_m \sigma N \tag{19}$$

where N is Avogadro's number. The standard values of σ are listed for each gaseous molecule in Reference 5; for N_2, $\sigma = 0.162$ nm².

2.2.4.2 Transmission method

In this method, the specific surface area is determined by measuring the transmission of fluid through a packed bed of powder. When high accuracy is not required, this method is frequently employed in industry because the apparatus is simple to operate and allows for quick measurement. The basis of this method is described by the Kozeny-Carman equation (Eq. 20). Letting U be the amount of fluid per unit time passing through the powder of cross-sectional area A and thickness L over which there exists a pressure difference ΔP, the weight-specific surface area S_w is given by:

$$S_w = (1/\rho)\left[g\Delta P \cdot A\varepsilon^3 / k\eta LU(1-\varepsilon)^2\right]^{1/2} \tag{20}$$

where ρ is the specific gravity of the powder, η the viscosity of fluid, ε the porosity of powder bed (i.e., $1 - (W/\rho LA)$), g is the acceleration of gravity, and k is a constant, called the Kozeny's constant is related to the porosity of the powder bed. The value of k is determined experimentally and is usually taken as 0.5.

The fluids used in the transmission method are liquid (such as water for powders consisting of large particles) or gaseous (mostly dried air, for powders of small particle size). In using this method, the amount of transmission U is measured at a constant pressure difference ΔP. The Blaine method is the simplest way for these measurements and instruments are commercially available. For these purposes, it is used generally for cement and other industrial powders according to JIS and ASTM.[6] The pressure difference ΔP, however, changes during measurement in this method. An apparatus for the transmission method using air and operating under constant pressure is shown in Figure 8 as an example. The principle and operation of the apparatus can be understood from the figure. The pressure difference ΔP is given by ρgh (ρ is the density of the liquid that produces the pressure difference) and can be varied over a range of 10 to 100 g cm^{-2} in an apparatus using water.

The most important factor in the transmission method is the packing of the powder. Eq. 20 takes into account the porosity of samples ε, but the dependence of the weight-specific surface area S_w on the porosity is, in fact, very complicated. Measurement of porosities will be described later (see 2.3.1).

2.2.5 Scattering of electromagnetic waves caused by particles

The optical properties of a particle depend on the state of assemblage of particles; a dilute dispersion of particles has different properties from thick powder aggregates such as those encountered in powder beds or in paints.[7] In the former state, the optical character of individual particles dominates, whereas in the latter, complicated multiple scattering occur. Multiple scattering is discussed elsewhere (see Chapter 3). On the other hand, X-ray scattering from particles is not affected by the assemblage of particles, but does depend on other properties such as the size and structure of particles.

2.2.5.1 Light scattering method

Light scattering depends largely on physical properties such as the shape of the particles, their mean diameter, and their physical state (i.e., liquid or solid). Typical changes in the scattering due to a change in the particle diameter are summarized in Table 6.[8] Theoretically, there are the Rayleigh scattering and the Mie equations; the latter equation extends Rayleigh scattering to larger particles. The Mie equation was obtained by solving electromagnetically the interaction between particles and light and is applicable to dispersed systems of powders. The scattered light intensity $I(\theta)$ in the direction θ to the incident light is given by:

$$I(\theta) = \lambda^2 \{i_1 + i_2(\theta)\} / 8\pi^2 R^2 \tag{21}$$

where λ is the wavelength, R is the distance between particle and observation point, i_1 is the light component having its electric vector perpendicular to the plane of observation

Chapter two: Measurements of powder characteristics

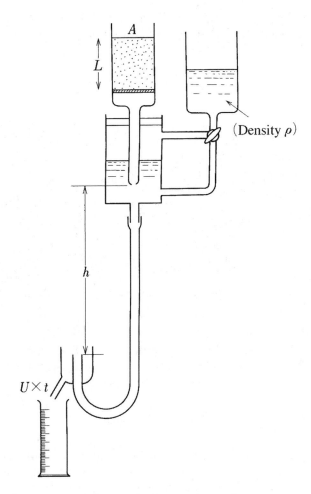

Figure 8 Apparatus for air transmission method, operating under constant pressure.

Table 6 Characteristics of Light Scattering Phenomena Depending on Particle Diameter

Particle diameter	x	Character of light scattering used for measurement	Characteristics
0–0.1 µm	0–0.5	Proportional to the square of volume and inversely proportional to wavelength to the 4th power	Rayleigh scattering
0.1–0.2 µm	0.5–1	1. Scattering angle for maximum polarization 2. Ratio of polarization in the direction of 90° 3. Color of scattering light in the direction of observation 4. Intensity ratio of scattering light at 45° and 135°	Mie scattering (3-term approximation)
0.2–2.0 µm	1–10	1. Change of absorption coefficient by wavelength 2. Maximum and minimum angles in scattering pattern 3. Tyndall spectrum of high order	Complicated diffraction and scattering regions
2–10 µm	10–50	1. Maximum and minimum sites in diffraction pattern 2. Diameter of diffraction ring and angle between rings	Fraunhofer diffraction
>10 µm	>50	1. Maximum site of rainbow 2. Color of shining ring 3. Shadowing by cross-section of particle	Geometrical optics region

Note: $x = \pi dm/\lambda$ (d: particle diameter, m: relative diffractive index of particle to medium, λ: wavelength).

and is independent of θ, and $i_2(θ)$ is the light component parallel to the plane and is dependent on θ. The total amount of scattered light S is given by:

$$S = (\lambda^2/2\pi)\sum_{n=1}^{\infty}(2n+1)(|a_n|^2 + |b_n|^2) \qquad (22)$$

where $|a_n|$ and $|b_n|$ are complicated functions that include Ricatti-Bessel functions and their derivatives. Both functions depend on θ and a parameter x (= $2\pi rm/\lambda$, where r is the radius of spherical particle and m the refractive index) and have been tabulated.[9,10] The radial distribution of the scattered light is also complicated, as is its wavelength dependence; interference occurs when the particle diameter approaches the wavelength of light. In actual measurements, one obtains the ratio of the scattered light at two different angles (0° and 90°, or 45° and 135°), and compares these with the calculated values through Eq. 22.

In practice, there are two methods that use light scattering. Light scattering photometers determine the mean particle diameter and the degree of dispersion by measuring the angular distribution of the scattered light from a large number of particles. In the other method, the particle-size distribution is obtained by measuring the scattered light intensity of individual particles.

2.2.5.2 Diffraction method

Measurements with this method are performed using Fraunhofer diffraction and are effective for particles having diameters around 1 to 10 μm. The intensity of the diffracted light due to a disk as a function of the angle θ is given as:

$$S(\theta) = x^2(1+\cos\theta)J_1(x\sin\theta)/2x\sin\theta \qquad (23)$$

where x is $2\pi rm/\lambda$ (r: the radius of particle, m: the relative refractive index) as mentioned above, and J_1 is a Bessel function. When x sinθ is larger than 10, J_1(x sinθ)/sinθ is negligible. Normalized values of S(θ) are shown in Figure 9.

In this method, an intense collimated light beam is made incident on a system of particles dispersed in a liquid or gaseous phase; the angular distribution of the scattered light intensity is measured. The particle-size distribution is determined computationally using the angular distribution as a function of x (Figure 9) in conjunction with Eq. 23. Several kinds of instruments for these measurements are commercially available; some commercial equipment is capable of yielding accuracies down to ±0.1 μm.

2.2.5.3 X-ray diffraction and X-ray scattering

The Debye-Scherrer method is used to obtain X-ray diffraction measurements in powders. It is well known that broadening occurs in the diffraction rings as the particle size of the powders decreases. The width at half height of the diffracted ray, as shown in Figure 10, is related to the particle diameter of the crystallite D by the Scherrer equation:

$$D = K\lambda/(B-b)\cos\theta \qquad (24)$$

where λ is the wavelength of monochromatic X-ray, θ is the Bragg angle (the diffracted ray appears at angle 2θ). b and B are the peak widths at half height for small and large (larger than ~10 μm) crystallites, respectively, as shown in the figure. This method is applicable to crystallites ranging from 1.0 to 0.01 μm in diameter, but the grains must have good crystallinity.

Chapter two: Measurements of powder characteristics 73

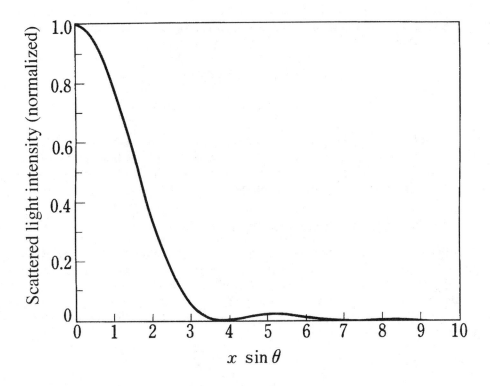

Figure 9 Intensity distribution of Fraunhofer diffraction.

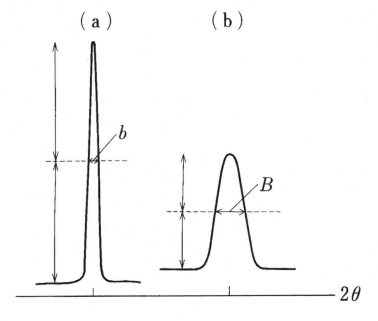

Figure 10 Intensity curves of X-ray diffraction profiles for large (a) and small particles (b).

The method for measuring the angular distribution of the intensity of the scattered X-ray is called the small angle method because only small incident angles are involved. The intensity of the scattered radiation is given approximately by:

$$\log I(\theta) = \log I_0 - \pi^2 D^2 s^2 / 3\lambda^2$$
$$s = (4\pi/\lambda)\sin(\theta/2) \sim 2\pi\theta/\lambda \tag{25}$$

A plot of log $I(\theta)$ vs. s^2 is linear and its slope gives the particle diameter. This method is applicable to particles ranging in size from 1 to 200 nm. Recently, instruments for measuring small angle scattering as well as line broadening have become commercially available as attachments to analytic X-ray equipment.

2.3 Measurements of packing and flow

The properties of each powder particle appear fairly distinctly when the powder is dispersed in a medium. Usually, however, aggregations of dried particles are found in a container or in a pile. One of the characteristics of dried powders is that they behave like a fluid, even though they are solid particles. The static and dynamic properties characteristic of powders depends on the mechanical interactions between individual particles.[10,11]

2.3.1 Definition of packing

The following quantities are used for describing the packing pattern of powder particles, i.e., the degree of packing, (See Figure 11):

1. *Apparent specific volume:* the volume that a powder of unit weight occupies when packed:

$$V_a = V/W \tag{26}$$

2. *Apparent density, bulk density:* the reciprocal of the apparent specific volume:

$$\rho_a = W/V \tag{27}$$

3. *Porosity, void ratio:* the ratio of the volume of void to the total volume of packed powder:

$$\varepsilon = V_v/V = 1 - V_s/V \tag{28}$$

4. *Packing ratio:* the ratio of the volume of the substantial solid part to the total volume of packed powder:

$$\phi = V_s/V = 1 - \varepsilon \tag{29}$$

Among the quantities above, the apparent specific volume, porosity, and void ratio are used frequently; the conversion between each quantity is simple. For powders with

Chapter two: Measurements of powder characteristics 75

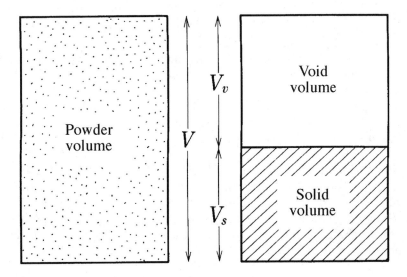

Figure 11 Expression for packing of powder (the weight of void is neglected).

secondary structures like porous powders, granules, and crushed powders of sintered materials, V_v is divided into a void volume inside the particles, V_1 and the void volume between particles, V_2. Then, $V = V_s + V_1 + V_2$. The solid density ρ_s and the grain density ρ_g (also called the green density) are defined as follows:

$$\rho_s = W/V_s, \qquad \rho_g = W/(V_s + V_1) \tag{30}$$

For a powder dispersed in water or a liquid medium, the apparent specific volume of the powder sedimented by gravity, called the sedimentation volume, sometimes also is used.

2.3.2 Measurements of apparent density

Methods for measuring the apparent density shown in JIS, etc. are described below. This density is usually measured using home-made apparatuses. According to the measuring methods, the following names are used.

1. *Static bulk density:* A powder sample is sieved, piled into a vessel, and its density is measured. This method gives the bulk density in the loosest packing state of the powder.
2. *Funnel-damper bulk density:* A powder sample is placed in a funnel and is transferred into a vessel by opening the funnel aperture quickly; the bulk density of the material in the vessel is then measured. This density is the loosely packed bulk density. This method is used frequently in JIS, but to make this method a standard, it is necessary to specify the shape and size of the funnel, the size of the aperture, and the position, shape, and size of the vessel.
3. *Lateral vibration bulk density:* This method packs the sample by producing lateral vibrations in a side of the vessel. Again, the bulk density measured is for a relatively loosely packed sample.

4. *Vertical vibration bulk density:* This method packs the sample by giving vertical vibrations to the vessel; this measurement gives a denser bulk density because it is a more efficient compaction method than the lateral vibration procedure.
5. *Tap-bulk density:* This is the bulk density after the sample is packed into a constant volume by tapping; tapping heights of powders are usually 1 or 2 cm. The powders in this case are considerably denser.
6. *Compression bulk density:* This is the bulk density of a cake-like powder obtained by compressing with a piston after the powder sample is placed in a cylindrical vessel. This method is similar to cold casting.

The bulk density of relatively large spherical particles (larger than about 100 μm) is not very dependent on the measurement method used; however, it is not unusual that for some powders, the ratio of the static bulk density (1) to the tap-bulk density (2) is used when the difference in packing states between samples is sought; the tap-bulk density (5) is suitable for discussing the relation between the bulk density and other physical properties of the compacted material.

2.3.3 Measurements of fluidity

The following experiments are carried out to measure the fluidity of powders through measurements of the tap-packing process, determination of the shearing stress inside a powder cake, rest angle, compressibility, and efflux rate from a hopper. These measurements, however, only look at one of the characters of the flow phenomena of powders. They are essentially all related to each other.

2.3.3.1 Rest angle

The measurement of the rest angle is widely used as a method to determine the fluidity of powder. This was one of the earliest measurements and continues to be used because it is a simple and convenient method; it is still widely accepted as a method for giving basic data. The rest angle is defined by ϕ in Figure 12; the mechanical meaning of the angle differs slightly, depending on the method of measurement.

In Figure 12, (a) and (b) give ϕ directly. In (c), the powder is placed into a horizontal cylinder and then the cylinder is slowly rotated. When the surface of the powder bed starts to slip, the inclination of the surface of the powder bed to a horizontal plane is measured, defining ϕ. (d) is much the same as (c), and only the vessel is different. The results obtained from all these methods depend on the shape and size of the apparatus; thus, the measurement is generally considered to be a relative one.

2.3.3.2 Motion angle

In determining the rest angle, the internal friction must be taken into consideration when the powder starts to slip; for the powder in motion, the dynamical friction of particles must also be considered. This friction is also important in determining the fluidity of powders. In the apparatus shown in Figure 12(c), the motion angle is given by the inclination angle Θ (different from ϕ) of the surface of the powder bed when the cylinder is rotating at a constant angular velocity ω. The viscosity coefficient of a fluid is proportional to $\sin \Theta / \omega$. By applying this relation to powders, it is possible to define a quantity corresponding to the viscosity coefficient.

2.3.3.3 Powder orifice

The efflux rate of powders from an orifice opened at the bottom of a vessel is a measure of the fluidity of powders and is also widely used. The difference from fluid flow is that the powder efflux rate does not depend significantly on the height of the powder. Various

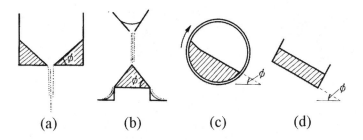

Figure 12 Measurement of the rest angle.

experimental relations between the weight efflux rate Q (kg/min) and the orifice diameter D_0 have been reported. The following equation, as an example, is applicable to the fluid orifice efflux rate:

$$Q = C(\pi/4)(g/2\mu)^{1/2} D_0^{5/2} \qquad (31)$$

where g is the acceleration of gravity, μ the frictional coefficient of the wall surface, and C the efflux coefficient to be experimentally determined.

References

1. Jimbo, G., *J. Soc. Powder Technology, Japan*, 17, 307, 1980 (in Japanese).
2. Masuda, M. and Iinoya, K., *J. Chem. Eng. Japan*, 4, 60, 1971.
3. Mullin, J.W. and Ang, H.M., *Powder Technol.*, 10, 153, 1974.
4. Alliet, D.F., *Powder Technol.*, 3, 3, 1976.
5. McClellan, A.L. and Harnsberger, H.F., *J. Colloid Interface Sci.*, 23, 577, 1967.
6. Homma, E. and Isono, N., *Annu. Rep. Cement Eng.*, XX, 135, 1966 (in Japanese).
7. Bohren, C.F. and Huffman, D.R., *Absorption and Scattering of Light by Small Particles*, Wiley-Interscience, 1983.
8. Hayakawa, S., *Surface and Fine Particles*, Kinoshita, K., Ed., Kyoritsu Publ. Co., 1986, 284 (in Japanese).
9. Kerker, M., *Scattering of Light and Other Electromagnetic Radiations*, Academic Press, 1963.
10. Hayakawa, S., *J. Japan Soc. Color Material*, 52, 515, 1979 (in Japanese).
11. Miwa, S., *Powder Science and Engineering*, 11(5), 44, 1979 (in Japanese).

chapter three — section one

Optical properties of powder layers

Kazuo Narita

Contents

3.1 Kubelka-Munk's theory 79
    3.1.1 Introduction 79
    3.1.2 Basic equations and their general solutions...80
    3.1.3 Light reflection and transmission of powder layers...82
    3.1.4 Optical properties of phosphor layers 84
        3.1.4.1 Cathode-ray excitation...84
        3.1.4.2 X-ray excitation 87
        3.1.4.3 Ultraviolet excitation 88
        3.1.4.4 Light output of fluorescent lamps 89
    3.1.5 Measurement of the scattering coefficient..91
References...92

3.1 Kubelka-Munk's theory

3.1.1 Introduction

Within a phosphor coating, excitation energy is absorbed after multiple scattering by individual particles.* The generated luminescent light also comes out of the coating after scattering. Hence, it is impossible to describe the total optical processes of absorption, excitation, and emission in a phosphor layer in a rigorously analytical manner, and some approximations must be introduced.

Three different approaches have been employed to calculate the optical properties of the phosphor layer. They are:

1. A method based on the solution of a set of simultaneous differential equations known as the Schuster-Kubelka-Munk equations, in which it is assumed that the phosphor layer is a continuous optical medium, and its optical properties are determined by two phenomenological constants, the absorption and scattering coefficients.

---

* In cases of excitation by X- and γ-rays, scattering can be neglected.

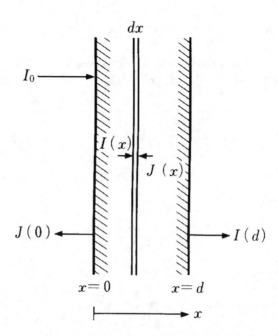

Figure 1 Absorption and scattering of light by a particle layer.

2. The Johnson method, in which the phosphor layer is regarded as a stack of thin layers, each layer having a thickness equal to the particle diameter of the phosphor; the optical properties of the entire layer is determined by an iterative calculation.
3. The Monte Carlo method, in which the path of the scattered light is determined in a stochastic manner using random numbers.

Each method has both advantages and disadvantages, and the most suitable method must be selected for each case. In this chapter, the details of these calculational methods and their applications to several kinds of phosphor coatings are presented.

3.1.2 Basic equations and their general solutions

Consider a layer consisting of a large number of particles and having a thickness d. It is assumed that light travels only in $\pm x$ directions in the layer, as shown in Figure 1.[1,2] Light with intensity I_0 falls on the layer, and is scattered in $+x$ and $-x$ directions with intensities $I(x)$ and $J(x)$, respectively. Let the absorption coefficient of the layer be k and the scattering coefficient be s. In the $+x$ direction, then, the light absorbed in an infinitesimal layer of thickness dx is $kI(x)dx$, and the light scattered is $sI(x)dx$. Further, a part of $J(x)$ (i.e., $sJ(x)dx$) is scattered back and has to be added to $I(x)$. Thus, one obtains:

$$\frac{dI(x)}{dx} = -(k+s)I(x) + sJ(x) \tag{1}$$

By considering the change of $J(x)$, one can write a similar equation:

$$\frac{dJ(x)}{dx} = (k+s)J(x) - sI(x) \tag{2}$$

Chapter three: Optical properties of powder layers

The general solutions to this set of equations are:

$$I(x) = A(1-\beta_0)e^{\alpha_0 x} + B(1+\beta_0)e^{-\alpha_0 x} \tag{3}$$

$$J(x) = A(1+\beta_0)e^{\alpha_0 x} + B(1-\beta_0)e^{-\alpha_0 x} \tag{4}$$

Here, A and B are constants whose values are determined by boundary conditions, and α_0 and β_0 are defined by:

$$\alpha_0 = \sqrt{k(k+2s)} \tag{5}$$

$$\beta_0 = \sqrt{k/(k+2s)} \tag{6}$$

Kubelka[3] generalized this one-dimensional formulation and showed that the same kind of equations can be derived even when the incident light is diffuse and scattering inside the particle layer takes place in all directions. In this case, the light is not always falling on the layer perpendicularly. Hence, the length of the light path in the layer is longer than dx. It can be verified that the mean path $d\xi$ is expressed as:

$$d\xi = 2dx \tag{7}$$

Therefore, Eq. 1 can be written as:

$$dI(x) = -(k+s)I(x) \cdot 2dx + sJ(x) \cdot 2dx \tag{8}$$

If one defines the new coefficients K and S by:

$$2k \equiv K, \quad 2s \equiv S \tag{9}$$

then, equations similar to Eqs. 1 and 2 are obtained:

$$\frac{dI(x)}{dx} = -(K+S)I(x) + SJ(x) \tag{10}$$

$$\frac{dJ(x)}{dx} = (K+S)J(x) - SI(x) \tag{11}$$

The solutions can be written down simply by replacing α_0 and β_0 in Eqs. 3 and 4 by α and β, where

$$\alpha = \sqrt{K(K+2S)} \tag{12}$$

$$\beta = \sqrt{K/(K+2S)} \tag{13}$$

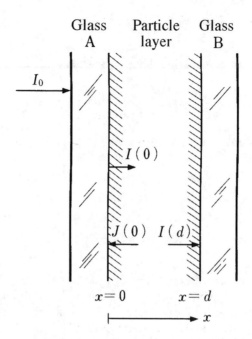

Figure 2 Particle layer between two glass plates.

The results are:

$$I(x) = A(1-\beta)e^{\alpha x} + B(1+\beta)e^{-\alpha x} \tag{14}$$

$$J(x) = A(1+\beta)e^{\alpha x} + B(1-\beta)e^{-\alpha x} \tag{15}$$

3.1.3 Light reflection and transmission of powder layers[4]

Consider a particle layer placed between glass plates A with reflectance r_a and transmittance t_a, and B with r_b and t_b, respectively (Figure 2). If the layer thickness is d, the boundary conditions are:

$$I(0) = t_a I_0 + r_a J(0) \tag{16}$$

$$J(d) = r_b I(d) \tag{17}$$

By introducing ρ_a and ρ_b defined by:

$$\rho_a = (1-r_a)/(1+r_a) \tag{18}$$

$$\rho_b = (1-r_b)/(1+r_b) \tag{19}$$

Eqs. 16 and 17 can then be written as:

$$(1+\rho_a)I(0) = t_a(1+\rho_a)I_0 + (1-\rho_a)J(0) \tag{20}$$

$$(1+\rho_b)J(d) = (1-\rho_b)I(d) \tag{21}$$

From the general solutions in Eq. 14, Eq. 15, and the above boundary conditions, the constants A and B are found to be:

$$A = -t_a I_0 \frac{(\rho_b - \beta)(1+\rho_a)e^{-\alpha d}}{2(\rho_a + \beta)(\rho_b + \beta)e^{\alpha d} - 2(\rho_a - \beta)(\rho_b - \beta)e^{-\alpha d}} \tag{22}$$

$$B = t_a I_0 \frac{(\rho_b + \beta)(1+\rho_a)e^{\alpha d}}{2(\rho_a + \beta)(\rho_b + \beta)e^{\alpha d} - 2(\rho_a - \beta)(\rho_b - \beta)e^{-\alpha d}} \tag{23}$$

If a glass plate exists only on one side (for example only glass B), then $t_a = 1$ and $r_a = 0$ ($\rho_a = 1$). In this case, transmittance T and reflectance R of the particle layer are given by Eqs. 24 and 25, respectively.

$$T = \frac{I(d)}{I_0} t_b = t_b \frac{(1+\beta)(\rho_b + \beta) - (1-\beta)(\rho_b - \beta)}{(1+\beta)(\rho_b + \beta)e^{\alpha d} - (1-\beta)(\rho_b - \beta)e^{-\alpha d}} \tag{24}$$

$$R = \frac{J(0)}{I_0} = \frac{(1-\beta)(\rho_b + \beta)e^{\alpha d} - (1+\beta)(\rho_b - \beta)e^{-\alpha d}}{(1+\beta)(\rho_b + \beta)e^{\alpha d} - (1-\beta)(\rho_b - \beta)e^{-\alpha d}} \tag{25}$$

Reflectance of a powder layer having a semi-infinite thickness, R_∞, is of practical importance, as reflection spectra of phosphors are measured in this manner. By extrapolating $d \to \infty$ in Eq. 25, one obtains:

$$R_\infty = \frac{1-\beta}{1+\beta} \tag{26}$$

Using Eq. 13, this can be rewritten as:

$$\frac{K}{S} = \frac{(1-R_\infty)^2}{2R_\infty} \equiv F(R_\infty) \tag{27}$$

The resultant function $F(R_\infty)$ is called the remission or Kubelka-Munk function. If S is independent of wavelength, $F(R_\infty)$ is proportional to the absorption coefficient K.

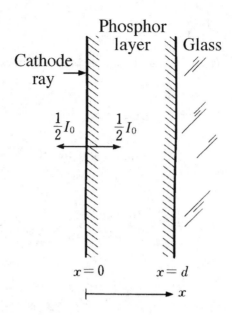

Figure 3 Generation of cathodoluminescence in the phosphor screen of a cathode-ray tube.

3.1.4 Optical properties of phosphor layers

The excitation energy given to a phosphor layer penetrates to the interior of the layer, with the intensity becoming gradually smaller because of absorption and scattering by the phosphor particles. Luminescence light arises approximately in proportion to the absorbed excitation energy, and emerges out of the layer after multiple scattering and some absorption. The entire process of excitation and luminescence emission by phosphor layers is described in the following.

3.1.4.1 Cathode-ray excitation

Bril et al.[5] calculated the intensity of cathodoluminescence emission on the assumption that the energy of incident electrons is completely absorbed at the surface of the phosphor layer.*

Figure 3 illustrates the face plate of a cathode-ray tube. If emission with intensity I_0 is generated at the surface of the phosphor coating, $I_0/2$ of it is emitted into space and another $I_0/2$ is directed into the inside of the layer. The boundary conditions for the emission of intensity $I(x)$ and $J(x)$ are:

$$I(0) = I_0/2 \quad (x = 0) \tag{28}$$

$$J(d) = r_g I(d) \quad (x = d) \tag{29}$$

where r_g is the reflectance of the glass.
When ρ_g is defined by:

$$\rho_g = \frac{1 - r_g}{1 + r_g} \tag{30}$$

---

* This approximation can be applied to all cases in which the penetration depth of the incident energy is much smaller than the thickness of the layer.

one can calculate the absorption and scattering of the emitted light using Eqs. 14 and 15. The result is:

$$I(d) = I_0 \frac{\beta(1+\rho_g)}{(1+\beta)(\rho_g+\beta)e^{\alpha d} - (1-\beta)(\rho_g-\beta)e^{-\alpha d}}$$

$$= \frac{1}{2} I_0 \frac{1}{\dfrac{K+(1-r_g)S}{\alpha}\sinh\alpha d + \cosh\alpha d}$$

(31)

$$J(0) = \frac{1}{2} I_0 \frac{(1-\beta)(\rho_g+\beta)e^{\alpha d} - (1+\beta)(\rho_g-\beta)e^{-\alpha d}}{(1+\beta)(\rho_g+\beta)e^{\alpha d} - (1-\beta)(\rho_g-\beta)e^{-\alpha d}}$$

$$= \frac{1}{2} I_0 \frac{r_g \cosh\alpha d + \dfrac{-Kr_g+(1-r_g)S}{\alpha}\sinh\alpha d}{\dfrac{K+(1-r_g)S}{\alpha}\sinh\alpha d + \cosh\alpha d}$$

(32)

The emission intensity observed at the excitation side, I_R, is given by:

$$I_R = J(0) + \frac{1}{2} I_0$$

(33)

while the emission intensity at the glass side, I_T, is:

$$I_T = t_g I(d)$$

(34)

where t_g is transmittance of the glass. If a metallic reflecting film of reflectance r_a is deposited on the phosphor layer, as is the practice for the commercial cathode-ray tube, the boundary condition

$$I(0) = \frac{I_0}{2} + \left(\frac{I_0}{2} + J(0)\right) r_a$$

(35)

is used in place of Eq. 28. In this case, the solution is given by Eq. 36 instead of Eq. 31,

$$I(d) = I_0 \frac{\beta(1+\rho_g)}{(\rho_a+\beta)(\rho_g+\beta)e^{\alpha d} - (\rho_a-\beta)(\rho_g-\beta)e^{-\alpha d}}$$

(36)

where ρ_a is defined as:

$$\rho_a = \frac{1-r_a}{1+r_a}$$

(37)

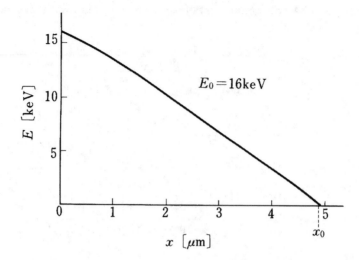

Figure 4 Relation between electron energy and penetration depth. (Phosphor: ZnCdS (60:40), acceleration voltage = 16 kV). (From Giakoumakis, G.E., Nomicos, C.D., and Euthymiou, P.C., *Can. J. Phys.*, 59, 88, 1981. With permission.)

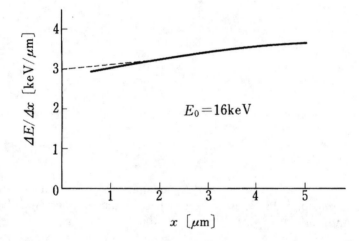

Figure 5 Relation between electron energy loss per unit length and penetration depth (Phosphor: ZnCdS (60:40), acceleration voltage = 16 kV). (From Giakoumakis, G.E., Nomicos, C.D., and Euthymiou, P.C., *Can. J. Phys.*, 59, 88, 1981. With permission.)

If the acceleration voltage is below 10 kV, the penetration depth of electrons into the phosphor layer, which is usually 20 to 30 µm thick, is estimated to be less than 2 µm. Hence, the above approximation is justified. When the acceleration voltage is greater, however, a more sophisticated treatment is necessary. Work by Giakoumakis[6] is described below.

A calculation based on Landau's theory on the energy loss of electrons in matter shows that an electron loses its energy in a phosphor layer [in this case, (Zn,Cd)S:Ag] in a manner shown in Figure 4. Differentiation of this curve yields the energy loss per unit thickness, $\Delta E/\Delta x$, which is shown in Figure 5. If the energy loss can be expressed by:

$$\frac{\Delta E}{\Delta x} = u + nx \qquad (u, n: \text{constants}) \tag{38}$$

and if the luminescence intensity is proportional to $\Delta E/\Delta x$, one obtains a pair of equations (Eqs. 39 and 40) on emission with intensities $I(x)$ and $J(x)$:

$$\frac{dI(x)}{dx} = -(K+S)I(x) + SJ(x) + \frac{1}{2}C_e(u+nx) \tag{39}$$

$$\frac{dJ(x)}{dx} = (K+S)J(x) - SI(x) - \frac{1}{2}C_e(u+nx) \tag{40}$$

where C_e = factor of proportionality between energy loss and emission intensity.

As Figure 4 shows, however, an electron penetrates only to a depth x_0, and Eqs. 39 and 40 are valid only for $0 \le x \le x_0$. When $x_0 \le x \le d$ (d = phosphor layer thickness), Eqs. 10 and 11 must be used.

By solving these four equations, the luminescence intensity radiated to the outside through the glass, I_T, is found to be:

$$I_T = t_g C_e (\rho_g + 1) \frac{(\rho_a - \beta)(P+n)e^{-\alpha x_0} + (\rho_a + \beta)(P-n)e^{\alpha x_0} + 2\beta\{n - u\rho_a(K+2S)\}}{2K(K+2S)\{(\rho_a + \beta)(\rho_g + \beta)e^{\alpha d} - (\rho_a - \beta)(\rho_g - \beta)e^{-\alpha d}\}} \tag{41}$$

where P is defined by:

$$P = \beta(K+2S)(u+nx_0) \tag{42}$$

3.1.4.2 X-ray excitation[4]

X-ray radiation penetrates the phosphor layer without being scattered, and its intensity decreases exponentially in accordance with Lambert's law. Luminescence of intensity $c_x e^{-\mu x} dx$ is generated in the layer of thickness dx, where c_x is constant and μ is the X-ray absorption coefficient of the phosphor. The following equations describe the emission intensity changes in the $+x$ and $-x$ directions.

$$\frac{dI(x)}{dx} = -(K+S)I(x) + SJ(x) + \frac{1}{2}c_x e^{-\mu x} \tag{43}$$

$$\frac{dJ(x)}{dx} = (K+S)J(x) - SI(x) - \frac{1}{2}c_x e^{-\mu x} \tag{44}$$

The general solutions of these equations are

$$I(x) = A(1-\beta)e^{\alpha x} + B(1+\beta)e^{-\alpha x} - \frac{1}{2}c_x \frac{\mu + \alpha/\beta}{\mu^2 - \alpha^2} e^{-\mu x} \tag{45}$$

$$J(x) = A(1+\beta)e^{\alpha x} + B(1-\beta)e^{-\alpha x} + \frac{1}{2}c_x \frac{\mu - \alpha/\beta}{\mu^2 - \alpha^2} e^{-\mu x} \tag{46}$$

For X-ray intensifying screens, the phosphor layer of the screen is placed between a photographic film and a plastic base, as shown in Figure 6. It is assumed that the photo-

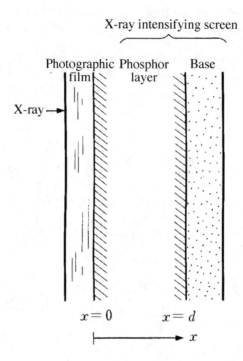

Figure 6 X-ray photography using an intensifying screen.

graphic film and the base have reflectances for the emitted light, r_f and r_c, respectively. Then, the boundary conditions are as follows:

$$I(0) = r_f J(0) \quad \text{or} \quad (1+\rho_f)I(0) = (1-\rho_f)J(0) \tag{47}$$

$$J(d) = r_c I(d) \quad \text{or} \quad (1+\rho_c)J(d) = (1-\rho_c)I(d) \tag{48}$$

The luminescence intensity at the photographic plate, $J(0)$, is given by Eq. 49.

$$J(0) = \frac{1}{2}c_x(1+\rho_f)\frac{(\rho_c+\beta)(\mu-\alpha)e^{\alpha d} - (\rho_c-\beta)(\mu+\alpha)e^{-\alpha d} + 2(\alpha\rho_c - \mu\beta)e^{-\mu d}}{\{(\rho_f+\beta)(\rho_c+\beta)e^{\alpha d} - (\rho_f-\beta)(\rho_c-\beta)e^{-\alpha d}\}(\mu^2-\alpha^2)} \tag{49}$$

3.1.4.3 Ultraviolet excitation[7–9]

The total process of photoluminescence is now described. Suppose a phosphor is coated on a glass plate, which is excited with ultraviolet radiation incident on the phosphor side. Absorption and scattering of the incident light can be described in the same manner as discussed in 3.1.3. The intensity of the light emitted by the phosphor layer of thickness dx is $qK(I(x) + J(x))dx$, where $I(x)$ and $J(x)$ are the intensity of ultraviolet radiation in the $+x$ and $-x$ directions, respectively, q is the energy efficiency of luminescence of the phosphor, and K is the absorption coefficient of the phosphor for ultraviolet radiation. Using Eqs. 14 and 15, the above intensity of light is found to be $2qK(Ae^{\alpha x} + Be^{-\alpha x})dx$. If the emitted light is directed equally to the $+x$ and $-x$ directions, the equations to describe its behavior are:

Chapter three: Optical properties of powder layers

$$\frac{dI'(x)}{dx} = -(K' + S')I'(x) + S'J'(x) + qK(Ae^{\alpha x} + Be^{-\alpha x}) \quad (50)$$

$$\frac{dJ'(x)}{dx} = (K' + S')J'(x) - S'I'(x) - qK(Ae^{\alpha x} + Be^{-\alpha x}) \quad (51)$$

To distinguish between exciting radiation and emitted light, symbols with a prime are used for the latter.

The general solutions to these equations are:

$$I'(x) = \frac{qKA}{\beta'} \cdot \frac{\alpha\beta' - \alpha'}{\alpha^2 - \alpha'^2} e^{\alpha x} - \frac{qKB}{\beta'} \cdot \frac{\alpha\beta' + \alpha'}{\alpha^2 - \alpha'^2} e^{-\alpha x} + A'(1 - \beta')e^{\alpha' x} + B'(1 + \beta')e^{-\alpha' x} \quad (52)$$

$$J'(x) = -\frac{qKA}{\beta'} \cdot \frac{\alpha\beta' + \alpha'}{\alpha^2 - \alpha'^2} e^{\alpha x} + \frac{qKB}{\beta'} \cdot \frac{\alpha\beta' - \alpha'}{\alpha^2 - \alpha'^2} e^{-\alpha x} + A'(1 + \beta')e^{\alpha' x} + B'(1 - \beta')e^{-\alpha' x} \quad (53)$$

Here α' and β' are defined by:

$$\alpha' = \sqrt{K'(K' + 2S')} \quad (54)$$

$$\beta' = \sqrt{K'/(K' + 2S')} \quad (55)$$

The values of A' and B' are calculated by the boundary conditions:

$$\begin{aligned} I'(0) &= 0 \\ J'(d) &= r'I'(d) \quad \text{or} \quad (1 + \rho')J'(d) - (1 - \rho')I'(d) = 0 \end{aligned} \quad (56)$$

with r' = reflectance of glass at the luminescence frequency

and are given by Eqs. 57 and 58.

$$A' = qK \frac{(1+\beta')\{A(\alpha\beta' + \alpha'\rho')e^{\alpha d} - B(\alpha\beta' - \alpha'\rho')e^{-\alpha d}\} + (\rho' - \beta')\{A(\alpha\beta' - \alpha') - B(\alpha\beta' + \alpha')\}e^{-\alpha' d}}{\beta'(\alpha^2 - \alpha'^2)\{(1+\beta')(\rho' + \beta')e^{\alpha' d} - (1-\beta')(\rho' - \beta')e^{-\alpha' d}\}} \quad (57)$$

$$B' = -qK \frac{(1-\beta')\{A(\alpha\beta' + \alpha'\rho')e^{\alpha d} - B(\alpha\beta' - \alpha'\rho')e^{-\alpha d}\} + (\rho' + \beta')\{A(\alpha\beta' - \alpha') - B(\alpha\beta' + \alpha')\}e^{\alpha' d}}{\beta'(\alpha^2 - \alpha'^2)\{(1+\beta')(\rho' + \beta')e^{\alpha' d} - (1-\beta')(\rho' - \beta')e^{-\alpha' d}\}} \quad (58)$$

3.1.4.4 Light output of fluorescent lamps

As shown in Figure 7(a), the ultraviolet radiation generated by the discharge in a fluorescent lamp is absorbed in the phosphor coating after multiple reflection. The total intensity of the ultraviolet radiation incident on the phosphor is:

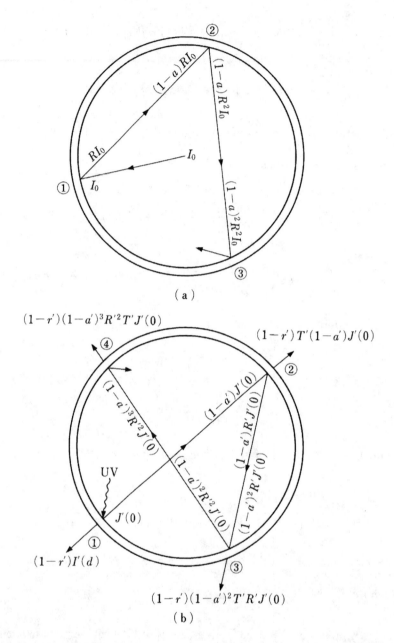

Figure 7 Multiple scattering of light inside a fluorescent lamp: (a) ultraviolet radiation; (b) luminescent light.

$$I_0 + (1-a)RI_0 + (1-a)^2 R^2 I_0 + \ldots = \frac{I_0}{1-(1-a)R} \tag{59}$$

where I_0 = intensity of ultraviolet radiation created by the discharge
a = proportion of ultraviolet radiation lost by absorption in the discharge space
R = reflectance of phosphor coating for ultraviolet radiation

Let the intensity of fluorescent light emitted to the excitation side be $J'(0)$, and that to the outer side be $I'(d)$. The emitted light is transmitted or reflected as shown in Figure 7(b), and leaves the lamp with intensity F given by:

$$F = \frac{1}{1-(1-a)R}\{(1-r')I'(d) + (1-r')(1-a')T'J'(0) + (1-r')(1-a')^2 T'R'J'(0) + \ldots\}$$

$$= \frac{1-r'}{1-(1-a)R}\left\{I'(d) + \frac{(1-a')T'}{1-(1-a')R'}J'(0)\right\}$$

(60)

where r' = reflectance of glass for the fluorescent light
         a' = proportion of fluorescent light lost by absorption in the discharge space and lamp ends
         R' = reflectance of phosphor layer for the fluorescent light
         T' = transmittance of phosphor layer for the fluorescent light

The values of $J'(0)$ and $I'(d)$ are calculated from Eqs. 52 and 53, and R, R', and T' from Eqs. 24 and 25.

3.1.5 Measurement of the scattering coefficient

To carry out the calculations mentioned above, a measured value of either the absorption or scattering coefficient must be known. If one of them is available, the other can be derived from Eq. 27 using the readily measurable reflectance of a thick layer. The easier way is to find the value of the scattering coefficient S by the reflectance measurements, described below, and then to calculate the absorption coefficient.

If one measures the reflectance of a sample layer R_0 coated on a black plate for which the reflectance can be neglected, the scattering coefficient S is given by[10-12]:

$$S = \frac{R_\infty}{d(1-R_\infty^2)} \ln \frac{R_\infty(1-R_0 R_\infty)}{R_\infty - R_0} \equiv \frac{1}{d}F(R_0, R_\infty)$$

(61)

where d is the thickness of the sample layer.

If Eq. 61 is valid, a d vs. $F(R_0, R_\infty)$ plot will yield a straight line, and its gradient gives the value of S. Figure 8 shows an experimental result on a calcium halophosphate phosphor.[13] It is seen that S is certainly a constant for this wavelength.

In general, the value of the scattering coefficient does not depend on wavelength when the particle size is sufficiently greater than the wavelength, typically >2 μ, while it is reciprocally proportional to particle sizes between 1 and 10 μ.[14-15] Further, in the case where absorption is not strong (i.e., $R_\infty \geq 0.50$), the scattering coefficient can be regarded as being independent of the absorption coefficient.[14]

In the ultraviolet region, where phosphors have strong absorption, it is impossible to use the above-mentioned method to determine S. The Kubelka-Munk theory itself does not predict the interdependence of S and K either. ter Vrugt[16] compared the Kubelka-Munk theory with Bodo's noncontinuous body approximation,[17] and concluded that the scattering coefficient is proportional to the absorption coefficient as far as $R_\infty \leq 0.30$. However, no experimental confirmation has been made. Therefore, application of the Kubelka-Munk theory to the ultraviolet region must be made carefully.

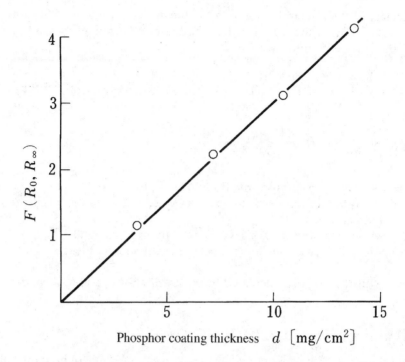

Figure 8 Relation between the coating thickness of phosphor and $F(R_0, R_\infty)$, measured at 500 nm. (From Narita, K., unpublished results, 1981. With permission.)

References

1. Schuster, A., *Astrophys. J.*, 21, 1, 1905.
2. Kubelka, P. and Munk, F., *Z. f. tech. Physik*, 12, 593, 1931.
3. Kubelka, P., *J. Opt. Soc. Am.*, 38, 448, 1948.
4. Hamaker, H.C., *Philips Res. Rep.*, 2, 55, 1947.
5. Bril, A. and Klasens, A., *Philips Tech. Rev.*, 15, 63, 1953.
6. Giakoumakis, G.E., Nomicos, C.D., and Euthymiou, P.C., *Can. J. Phys.*, 59, 88, 1981.
7. Ivanov, A.P., *Zhurnal eksper. teor. Fiz.*, 26, 275, 1954.
8. ter Vrugt, J.W., Manders, L.W.J., and Wanmaker, W.L., *Lighting Res. Technol.*, 7, 23, 1975.
9. Fonger, W.H., *Appl. Opt.*, 21, 1219, 1982.
10. Kortüm, G., *Reflectance Spectroscopy*, Springer, 1969, 112.
11. Gillespie, J.B., Lindberg, J.D., and Laude, L.S., *Appl. Opt.*, 14, 807, 1975.
12. Patterson, E.M., Shelden, C.E., and Stockton, B.H., *Appl. Opt.*, 16, 729, 1977.
13. Narita, K., unpublished data, 1981. Refer also to Reference 15.
14. Kortüm, G., Braun, W., and Herzog, G., *Angew. Chem.*, 75, 653, 1963.
15. Narita, K., *Electrochem. Soc. Spring Meeting Abstr.*, 1981, 177.
16. ter Vrugt, J.W., *Philips Res. Rep.*, 20, 23, 1965.
17. Bodo, Z., *Acta Physica Hungarica*, 1, 135, 1951.

chapter three — section two

Optical properties of powder layers

Akira Tomonaga* and Yoshiharu Komine

Contents

3.2 Johnson's theory 93
    3.2.1 Reflection, absorption, and transmission in a single layer 93
    3.2.2 Reflection, absorption, transmission and emission
          in a single layer within multiple layers 94
    3.2.3 Light flux of a fluorescent lamp 98
    3.2.4 Application to the white fluorescent lamp 99
References 103

3.2 Johnson's theory

As an application of Johnson's theory, this section presents an analysis of the relationship between the light flux of a fluorescent lamp and the thickness of its phosphor coating.

3.2.1 Reflection, absorption, and transmission in a single layer

First, assume that the thickness of a single phosphor layer is the same as the volume surface diameter d. The reflection, absorption, and transmission of incident light with an intensity I_0 are shown in Figure 9. In the figure, m indicates the ratio of the light reflected on the layer interface, and γ is the absorption coefficient (cm^{-1}) of the phosphor.

As shown in Figure 9, the incident light I_0 repeats the reflection, absorption, and transmission processes within the layer, and therefore, the final total reflection is:

$$I_0 m + I_0 m(1-m)^2 e^{-2\gamma d} + I_0 m^3(1-m)^2 e^{-4\gamma d} + \ldots$$

$$= I_0 m \left\{ 1 + \frac{(1-m)^2 e^{-2\gamma d}}{1 - m^2 e^{-2\gamma d}} \right\} \tag{62}$$

$$= I_0 \cdot t$$

---

* Akira Tomonaga is deceased.

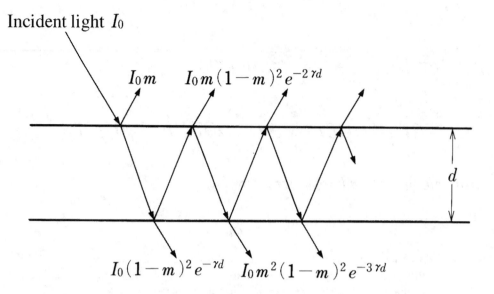

Figure 9 Reflectance, absorption, and transmission in a single phosphor layer.

The total transmission is:

$$I_0(1-m)^2 e^{-\gamma d} + I_0 m^2 (1-m)^2 e^{-3\gamma d} + \ldots$$

$$= I_0 \cdot \frac{(1-m)^2 e^{-\gamma d}}{1 - m^2 e^{-2\gamma d}} \qquad (63)$$

$$= I_0 \cdot s$$

Here, t and s are given by

$$t = m \left\{ 1 + \frac{(1-m)^2 e^{-2\gamma d}}{1 - m^2 e^{-2\gamma d}} \right\}$$

$$s = \frac{(1-m)^2 e^{-\gamma d}}{1 - m^2 e^{-2\gamma d}} \qquad (64)$$

Then, the absorbed light A_b is:

$$A_b = I_0 \cdot (1 - s - t) \qquad (65)$$

3.2.2 Reflection, absorption, transmission and emission in a single layer within multiple layers

Figure 10(a) shows a configuration of layers, and Figure 10(b) shows the reflection, absorption, and transmission that takes place in the ith layer. In figure 10(b), if the light traveling down (to the glass) and up (to the positive column) from the ith layer are denoted by I_i and J_i, respectively, the following equations hold:

Chapter three: Optical properties of powder layers

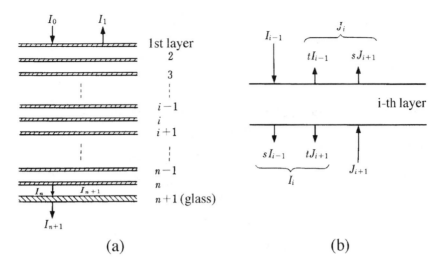

Figure 10 Layer configuration.

$$I_i = sI_{i-1} + tJ_{i+1}$$
$$J_i = sJ_{i+1} + tI_{i-1}$$
(66)

In Figure 10(a), if the (n + 1)th layer represents the glass, and s and t for the glass is denoted by s_g and t_g, respectively, then Eq. 67 gives the boundary conditions:

$$I_{n+1} = s_g I_n$$
$$J_{n+1} = t_g I_n$$
(67)

From these equations and those in Eq. 66, one obtains:

$$I_{n-i} = A_i I_n$$
$$J_{n-i+1} = B_i J_{n+1}$$
(68)

$$A_i = \frac{1}{s} A_{i-1} - \frac{t \cdot t_g}{s} B_{i-1}$$

$$B_i = \frac{t}{s \cdot t_g} A_{i-1} + \frac{s^2 - t^2}{s} B_{i-1}$$
(69)

$$A_0 = B_0 = 1$$

If it is assumed that:

$$\frac{1}{s} = \alpha$$

$$\frac{-t \cdot t_g}{s} = \beta$$

$$\frac{t}{s \cdot t_g} = \eta$$

$$\frac{s^2 - t^2}{s} = \delta \qquad (70)$$

then A_i and B_i can be represented using the matrix representation as follows:

$$\begin{pmatrix} A_i \\ B_i \end{pmatrix} = \begin{pmatrix} \alpha & \beta \\ \eta & \delta \end{pmatrix} \begin{pmatrix} A_{i-1} \\ B_{i-1} \end{pmatrix}$$

$$= \begin{pmatrix} \alpha & \beta \\ \eta & \delta \end{pmatrix}^i \begin{pmatrix} A_0 \\ B_0 \end{pmatrix} \qquad (71)$$

$$= \begin{pmatrix} \alpha_i & \beta_i \\ \eta_i & \delta_i \end{pmatrix} \begin{pmatrix} 1 \\ 1 \end{pmatrix}$$

Since $\alpha\delta - \beta\eta = 1$, α_i, β_i, η_i, and δ_i are expressed as follows:

$$\alpha_i = T_i(\theta) + \xi U_i(\theta)$$

$$\beta_i = \beta U_i(\theta)$$

$$\eta_i = \eta U_i(\theta) \qquad (72)$$

$$\delta_i = T_i(\theta) - \xi U_i(\theta)$$

$T_i(\theta)$ and $U_i(\theta)$ are *Chebyshev polynomials* expressed by Eq. 73.[1]

$$T_i(\theta) = \frac{1}{2}\left\{\left(\theta + \sqrt{\theta^2 - 1}\right)^i + \left(\theta - \sqrt{\theta^2 - 1}\right)^i\right\}$$

$$U_i(\theta) = \frac{1}{2} \cdot \frac{\left(\theta + \sqrt{\theta^2 - 1}\right)^i - \left(\theta - \sqrt{\theta^2 - 1}\right)^i}{\sqrt{\theta^2 - 1}} \qquad (73)$$

$$\theta = \frac{\alpha + \beta}{2}, \quad \xi = \frac{\alpha - \beta}{2}$$

From Eqs. 71, 72, and 73, one obtains Eq. 74.

$$A_i = \frac{1}{2}\left\{\left(1 + \frac{\beta + \xi}{\sqrt{\theta^2 - 1}}\right)\left(\theta + \sqrt{\theta^2 - 1}\right)^i + \left(1 - \frac{\beta + \xi}{\sqrt{\theta^2 - 1}}\right)\left(\theta - \sqrt{\theta^2 - 1}\right)^i\right\}$$

$$B_i = \frac{1}{2}\left\{\left(1 + \frac{\eta - \xi}{\sqrt{\theta^2 - 1}}\right)\left(\theta + \sqrt{\theta^2 - 1}\right)^i + \left(1 - \frac{\eta - \xi}{\sqrt{\theta^2 - 1}}\right)\left(\theta - \sqrt{\theta^2 - 1}\right)^i\right\} \qquad (74)$$

Chapter three: Optical properties of powder layers 97

Therefore, if s, t, s_g, and t_g are given, Eq. 68 can be used to obtain the relationship between the incident light and emitted light, in which the reflection between the layers is taken into account.

In Figure 10(b) the amount of light absorbed by the ith layer, A_{bi} is:

$$A_{bi} = I_{i-1} + J_{i+1} - I_i - J_i$$
$$= \frac{1-s-t}{s+t}(I_i + J_i) \tag{75}$$
$$= \frac{1-s-t}{s+t} \cdot \frac{1}{A_n}(A_{n-i} + t_g \cdot B_{n-i+1})I_0$$

To obtain Eq. 75, I_i can be expressed by substituting $(n - i)$ into Eq. 68 for i as follows:

$$I_i = A_{n-i}I_n \tag{76}$$

Similarly, J_i can be expressed by substituting $(n - i +1)$ for i in Eq. 68:

$$J_i = B_{n-i+1}J_{n+1}$$
$$= t_g B_{n-i+1}I_n \tag{77}$$

Using Eqs. 76 and 77, one has:

$$I_i + J_i = (A_{n-i} + t_g B_{n-i+1})I_n \tag{78}$$

By substituting n for i in Eq. 68,

$$I_0 = A_n I_n \tag{79}$$

Combining Eqs. 78 and 79, one obtains Eq. 75.

In the following, the ultraviolet and visible regions (fluorescence) are treated separately, so that these regions are indicated by adding the superscripts uv and v to the characters. Under this nomenclature, A_{bi}^{uv} stands for the ultraviolet rays absorbed by the ith layer. If the conversion efficiency to visible light is μ, then the visible light emitted from the ith layer is $2P = 1/2 \cdot \mu A_{bi}^{uv}$. One half, $P = 1/2 \cdot \mu A_{bi}^{uv}$, travels into the fluorescent lamp. The other half, $P = 1/2 \cdot \mu A_{bi}^{uv}$, escapes to the glass. As for the light traveling outward from the ith layer, P:

$$I_{n-i+1} = \frac{s_g^v}{A_{n-i}^v} \cdot P \tag{80}$$

is emitted through the glass to the outside. Some of the light P is reflected on the glass surface (reflection: $J_{n-i+1} = (t_g^v/A_{n-i}^v) \cdot P$) and

$$\frac{J_{n-i+1}}{A_n^v} = \frac{t_g^v}{A_n^v \cdot A_{n-i}^v} \tag{81}$$

is returned to the positive ion discharge column within the lamp structure. On the other hand, of the light P traveling inward from the ith layer, P/A_{i-1}^v is emitted to the positive ion discharge column. Therefore, the light E_i emitted from the ith layer to the outside of the lamp is:

$$E_i = \frac{s_g^v}{A_{n-i}^v} \cdot P \qquad (82)$$

and the light F_i emitted from the same layer to the positive ion discharge column is:

$$F_i = \left(\frac{1}{A_{i-1}^v} + \frac{t_g^v}{A_n^v \cdot A_{n-i}^v} \right) \cdot P \qquad (83)$$

Some of F_i is absorbed by the positive ion discharge column, and a fraction, ρ_v ($\rho^v \leq 1$), reenters the phosphor layer. Some of this light is again emitted to the outside of the lamp, and some is returned to positive ion discharge and partly absorbed by the column. This process is repeated several times so that the final emitted light to the outside of the lamp is:

$$G_i = \frac{s_g^v \cdot \rho^v}{A_n^v - B_n^v \cdot t_g^v \cdot \rho^v} \cdot F_i \qquad (84)$$

Therefore, the total visible light emitted from the ith layer to the outside of the lamp is expressed by the sum $(E_i + G_i)$.

3.2.3 Light flux of a fluorescent lamp

As shown in the previous section, the visible light emitted from the ith layer to the outside of the lamp is expressed by $(E_i + G_i)$. Therefore, the light emitted from the entire phosphor layer is $\Sigma(E_i + G_i)$. Here, it should be noted, however, that, just like the visible light, the incident ultraviolet light I_0 undergoes multiple reflection in the lamp, a part of it being absorbed by the positive column. For this reason, the effective incident ultraviolet light becomes:

$$\frac{I_0}{1 - \frac{B_n^{uv}}{A_n^{uv}} \cdot t_g^{uv} \cdot \rho^{uv}} \qquad (85)$$

Therefore, the visible light L_{pn} emitted from the phosphor multilayers to the outside of the lamp is:

$$L_{pn} = \frac{\sum_{i=1}^{n}(E_i + G_i)}{1 - \frac{B_n^{uv}}{A_n^{uv}} \cdot t_g^{uv} \cdot \rho^{uv}} \qquad (86)$$

Chapter three: Optical properties of powder layers

Furthermore, the visible light from the lamp contains mercury resonance lines which are transmitted:

$$L_{hn} = k \cdot \frac{s_g^v}{A_n^v - B_n^v \cdot t_g^v \cdot \rho^v} \cdot I_0 \quad (k = \text{constant}) \tag{87}$$

and the total light flux L_n from the fluorescent lamp is:

$$L_n = L_{pn} + L_{hn} \tag{88}$$

3.2.4 Application to the white fluorescent lamp

How well Eq. 88 agrees with measurements for a white fluorescent lamp using a white-light-emitting calcium halophosphate phosphor (volume surface diameter $d = 6.7$ μm)[2] is examined here.

According to Johnson,[3] the ratio m of reflection by one layer is $m = 1.5(n-1)^2/(n+1)^2$, where n is the refractive index and depends on the wavelength. The refractive indices for the visible and ultraviolet regions are represented by those of the Na D line and the Hg 254-nm-wavelength light. The refractive indices are 1.648 and 1.71 for the halophosphate phosphor, and 1.5 and 1.55 for glass.[2] Since the ratio m for the glass can be assumed to be $m = (n-1)^2/(n+1)^2$, the values of m for the phosphor and glass are as follows:

Halophosphate phosphor : $m^v = 0.090$, $m^{uv} = 0.103$

Glass : $m^v = 0.04$, $m^{uv} = 0.05$

The absorption coefficient γ of the halophosphate phosphor can be obtained by measuring the diffuse reflectance R of the phosphor with a large thickness. If the number of layers is increased to infinity in Figure 10, the diffuse reflectance is given by:

$$R = \lim_{n \to \infty} \frac{J_1}{I_0}$$

$$= t_g \lim_{n \to \infty} \frac{B_n}{A_n} \tag{89}$$

$$= t_g \cdot \frac{1 + \dfrac{\eta - \xi}{\sqrt{\theta^2 - 1}}}{1 + \dfrac{\beta + \xi}{\sqrt{\theta^2 - 1}}}$$

By substituting s, t, s_g, and t_g for β, η, θ, and ξ, Eq. 89 becomes:

$$R + \frac{1}{R} = \frac{1 - s^2 + t^2}{t} \tag{90}$$

Since $e^{-\gamma d}$ is obtained from Eqs. 90 and 64, and d is known, the coefficient γ can be determined. For practical purposes, however, Eq. 64 can be simplified by neglecting m^2, since m is much smaller than 1, as stated above.

$$s \simeq (1-2m)e^{-\gamma d}$$

$$t \simeq m\left\{1+(1-2m)e^{-2\gamma d}\right\} \quad (91)$$

In Eq. 90, by neglecting t^2 ($\ll 1$) and using Eq. 91, one obtains Eq. 92.

$$e^{-2\gamma d} = \frac{1-m\left(R+\dfrac{1}{R}\right)}{(1-2m)\left\{(1-2m)+m\left(R+\dfrac{1}{R}\right)\right\}} \quad (92)$$

This indicates that the absorption coefficient γ can be obtained by measuring the diffuse reflectance R. Since measurements of R give $R^{uv} = 0.18$ and $R^v = 0.86$, the absorption coefficient of the phosphor is $\gamma^{uv} = 750$ cm^{-1} and $\gamma^v = 3.6$ cm^{-1}. Therefore, the values of s and t for the phosphor can be obtained using Eq. 91.

Consider the values of s and t for the glass. By substituting measured values of $s_g^v = 0.92$–0.94 and $s_g^{uv} = 0$ into Eq. 91, one obtains $\gamma_g^v = 0$ and $\gamma_g^{uv} \simeq \infty$. These results can be used to obtain the values of t for the glass. Values of s and t for the phosphor and glass are listed in Table 1.

When the values of s and t are given the values of α, β, η, θ, and ξ in Eqs. 70 and 73 can be obtained; in turn, one can calculate the values of A_i and B_i in Eq. 74. The results are listed in Table 2.

With A_i and B_i, $(E_i + G_i)$ in Eq. 86 can be expressed by Eq. 93, and Eq. 86 becomes Eq. 94.

$$E_i + G_i = \frac{s_g^v}{2}\cdot\mu I_0 \cdot \frac{s^{uv}+t^{uv}}{1-s^{uv}-t^{uv}}\cdot \frac{A_{n-i}^{uv}+t_g^{uv}B_{n-i+1}^{uv}}{A_n^{uv}}\cdot\left\{\frac{1}{A_{n-i}^v}+\frac{\rho^v}{A_n^v-B_n^v\cdot t_g^v\cdot\rho^v}\left(\frac{1}{A_{i-1}^v}+\frac{t_g^v}{A_n^v\cdot A_{n-i}^v}\right)\right\} \quad (93)$$

$$L_{pn} = \frac{s_g^v}{2}\cdot\mu I_0 \cdot \frac{s^{uv}+t^{uv}}{1-s^{uv}-t^{uv}}\cdot \frac{1}{A_n^{uv}-B_n^{uv}\cdot t_g^{uv}\cdot\rho^{uv}}\cdot\left[\left\{1+\frac{\rho^v\cdot t_g^v}{A_n^v(A_n^v-B_n^v\cdot t_g^v\cdot\rho^v)}\right\}\sum_{i=1}^{n}\left(\frac{A_{n-i}^{uv}}{A_{n-i}^v}\right)\right.$$

$$+t_g^{uv}\left\{1+\frac{\rho^v\cdot t_g^v}{A_n^v(A_n^v-B_n^v\cdot t_g^v\cdot\rho^v)}\right\}\sum_{i=1}^{n}\left(\frac{B_{n-i+1}^{uv}}{A_{n-i}^v}\right)+\frac{\rho^v}{A_n^v-B_n^v\cdot t_g^v\cdot\rho^v}\sum_{i=1}^{n}\left(\frac{B_{n-1}^{uv}}{A_{i-1}^v}\right) \quad (94)$$

$$\left.+\frac{\rho^v\cdot t_g^{uv}}{A_n^v-B_n^v\cdot t_g^v\cdot\rho^v}\sum_{i=1}^{n}\left(\frac{B_{n-i+1}^{uv}}{A_{i-1}^v}\right)\right]$$

The results of calculations for $n = 1$ to 10 are shown in Table 3. The calculations are based on the values $\rho^v = 1.0$ and $\rho^{uv} = 0.65$.[4,5]

Next, using the same phosphor, prototypical 40-W fluorescent lamps are fabricated with varying coating weights of the phosphor to examine the relationship between the coating weight of the phosphor, w (g cm^{-2}), and the light flux (relative value). The results are shown in Table 4. In the table, the number of layers n was obtained using a halophosphate phosphor density of 3.2 g cm^{-3} and particle diameter (layer thickness) of 6.7×10^{-4} cm from Eq. 95.

Chapter three: Optical properties of powder layers 101

Table 1 Values of s and t for the Ultraviolet and Visible Regions

	Calcium halophosphate phosphor		Glass	
	s	t	s_g	t_g
Ultraviolet region	0.483	0.133	0	0.05
Visible region	0.818	0.164	0.92	0.08

Table 2 Values of A_i and B_i for the Ultraviolet and Visible Regions

	A_i	B_i
Ultraviolet region	$1.0245 \cdot 2.019^i - 0.0245 \cdot 0.495^i$	$3.58 \cdot 2.019^i - 12.25 \cdot 0.9145^i$
Visible region	$1.635 \cdot 1.0935^i - 0.635 \cdot 0.9145^i$	$13.25 \cdot 1.0935^i - 12.25 \cdot 0.9145^i$

Table 3 Calculated Results

n	A_n^{uv}	B_n^{uv}	A_n^v	B_n^v	$\sum_{i=1}^{n} \dfrac{A_{n-i}^{uv}}{A_{n-1}^v}$	$\sum_{i=1}^{n} \dfrac{B_{n-i+1}^{uv}}{A_{n-1}^v}$	$\sum_{i=1}^{n} \dfrac{A_{n-i}^{uv}}{A_{i-1}^v}$	$\sum_{i=1}^{n} \dfrac{B_{n-i+1}^{uv}}{A_{i-1}^v}$
1	2.056	5.951	1.207	3.286	1	5.951	1	5.951
2	4.170	13.96	1.424	5.598	2.703	17.51	2.542	18.88
3	8.429	29.15	1.652	7.955	5.632	37.98	6.575	44.89
4	17.02	59.33	1.894	10.38	10.73	73.89	13.93	96.88
5	34.37	120.0	2.150	12.88	19.72	137.3	28.71	201.2
6	69.40	242.5	2.424	15.49	35.71	250.0	58.47	411.3
7	140.1	489.6	2.717	18.22	64.34	452.0	118.5	835.0
8	282.9	988.5	3.032	21.09	115.9	815.8	239.6	1690
9	571.1	1996	3.371	24.14	20.92	1474	484.1	3415
10	1153	4030	3.737	27.38	378.6	2670	977.9	6899

n	$1 + \dfrac{\rho^v \cdot t_g^v}{A_n^v \left(A_n^v - B_n^v \cdot t_g^v \cdot \rho^v \right)}$	$\dfrac{\rho^v}{A_n^v - B_n^v \cdot t_g^v \cdot \rho^v}$	$\dfrac{1}{A_n^{uv} - B_n^{uv} \cdot t_g^{uv} \cdot \rho^{uv}}$	$\dfrac{L_{pn}}{\mu I_0}$
1	1.0701	1.0583	5.368×10^{-1}	1.094
2	1.0576	1.0245	2.691×10^{-1}	1.461
3	1.0477	0.9844	1.337×10^{-1}	1.635
4	1.0397	0.9404	6.625×10^{-2}	1.597
5	1.0332	0.8931	3.282×10^{-2}	1.504
6	1.0279	0.8445	1.625×10^{-2}	1.395
7	1.0234	0.7940	8.052×10^{-3}	1.285
8	1.0196	0.7439	3.988×10^{-3}	1.180
9	1.0165	0.6945	1.975×10^{-3}	1.082
10	1.0138	0.6465	9.784×10^{-4}	0.993

$$n = \frac{w}{3.2 \times 6.7 \times 10^{-4}} \tag{95}$$

$$= 4.7 \times 10^2 \cdot w$$

In Table 4, the light flux for $w = 0$ represents the light flux in the visible region of the mercury arc. Therefore, since Eq. 87 with $n = 0$ is:

Table 4 Light Flux Measurements (Relative Values) for Varying Coating Weights of Phosphor

	Measurement	
$w \times 10^3$	Light flux (relative value)	Number of layers (n)
0	0.091	0
0.66	0.305	0.31
1.32	0.495	0.62
1.98	0.817	0.92
3.43	0.958	1.6
5.85	0.998	2.7
10.3	0.916	4.8
16.7	0.776	7.8

Table 5 L_n and the Number of Layers (Calculated Values)

n	L_{hn}	L_{pn}	L_n
0	0.091	0	0.091
1	0.089	0.614	0.703
2	0.086	0.820	0.906
3	0.083	0.917	1.000
4	0.079	0.896	0.975
5	0.075	0.834	0.909
6	0.071	0.783	0.854
7	0.067	0.721	0.788
8	0.062	0.662	0.724
9	0.058	0.607	0.665
10	0.054	0.557	0.611

$$L_{ho} = k \frac{0.92}{1 - 0.08} I_0$$

$$= kI_0 \tag{96}$$

$$= 0.091$$

one obtains:

$$L_{hn} = \frac{0.084}{A_n^v - 0.08 B_n^v} \tag{97}$$

Table 5 shows the calculated dependence of L_{hn} on the number of layers n. In Table 3, $L_{pn}/\mu I_0$ reaches a maximum at $n = 3$. Thus, if one assumes that:

$$L_{p3} + L_{h3} = L_{p3} + 0.083$$
$$= 1 \tag{98}$$

then, $L_{p3} = 0.917$. Therefore, from:

$$\frac{L_{p3}}{\mu I_0} = \frac{0.917}{\mu I_0} = 1.635 \tag{99}$$

Chapter three: Optical properties of powder layers

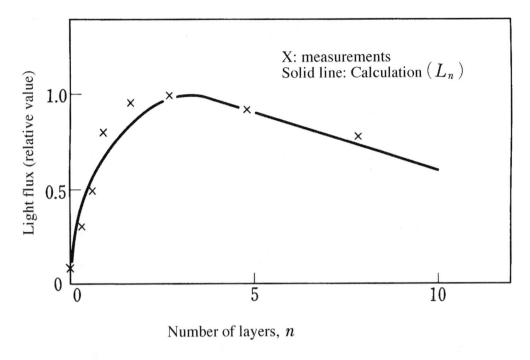

Figure 11 Relation between the number of layers and light flux. (From Tomonaga, A., Sueyasu, T., and Komine, Y., *Mitsubishi Denki Giho*, 46, 416, 1972 (in Japanese). With permission.)

$\mu I_0 = 0.561$; values of L_{pn} are shown in Table 5 as well as those for the light flux, L_n ($= L_{pn} + L_{hn}$), plotted in Figure 11. Figure 11 shows measured values for comparison. The figure reveals that the measurement and calculation agree almost completely for $n \geq 3$, whereas they do not agree very well for $n \leq 2$. Particularly for $1 \leq n \leq 2$, the measurements are larger than the calculated values. The numbers of layers giving the maximum light flux are 2.2 for the experimental and 3.0 for the calculated values. The most probable reasons for these discrepancies are:

1. The values of the absorption coefficient, s, and t were calculated using approximate equations.
2. There may be a difference in the thickness of the phosphor layer within the lamp.
3. Pores within the phosphor layer composed of phosphor particles were disregarded.
4. Real phosphor powder does not have a uniform particle size. However, the theory omits the distribution of particle sizes.

Despite these problems, Johnson's theory, which employs easily conceivable parameters such as particle diameter, number of layers, and absorption coefficient, is an effective method for analyzing the phosphor thickness of fluorescent lamps.

References

1. Kawakami, M., *Kisodenkikairo (Basic Electrical Circuit)*, Koronasha, 1960 (in Japanese).
2. Tomonaga, A., Sueyasu, T., and Komine, Y., *Mitsubishi Denki Giho*, 46, 416, 1972 (in Japanese).
3. Johnson, P.D., *J. Opt. Soc. Am.*, 42, 978, 1952.
4. Bo, H. and Takeyama, S., *J. Illum. Eng. Inst. Japan*, 44, 227, 1960 (in Japanese).
5. Ouweltjes, J.L., *Electrizitas Verwertung*, 11, 12, 1958.

chapter three — section three

Optical properties of powder layers

Koichi Urabe

Contents

3.3 Monte Carlo method 105
    3.3.1 Features of the Monte Carlo method 105
    3.3.2 Monte Carlo simulations for phosphor screens 105
        3.3.2.1 Particle layer model..106
        3.3.2.2 Continuum model..110
    3.3.3 Further problems 113
References...113

3.3 Monte Carlo method

3.3.1 Features of the Monte Carlo method

Problems of light scattering and absorption in powder layers are easily handled by two-flux methods based on Kubelka-Munk's theory or Johnson's theory when the path of light can be reduced to one dimension. Calculations for powder screens with a complicated structure, however, or those for the resolution of simple and uniform screens require a method that explicitly uses tilted light rays. For example, it is difficult to use a two-flux method to find out how much emitted light is absorbed by each component in the phosphor screen of a color picture tube like that shown in Figure 12. A six-flux method[1–4] or a many-flux method[5] is expected to give better results in some cases, but such methods are still not adequate for these types of practical problems.

    Several authors have used Monte Carlo methods to represent the angular dependence of light intensities.[6–9] A Monte Carlo method is a mathematical method explicitly utilizing random numbers and is applicable to any problems that can be formulated probabilistically. This method is also compatible with various boundary conditions, and it seems that there is no simpler method meeting the above requirement.

3.3.2 Monte Carlo simulations for phosphor screens

Soules[6] studied the light scattering of powder layers composed of spherical or randomly shaped particles. By using a Monte Carlo method, scattering parameters to simulate the angular distribution of reflected and transmitted light were obtained. This seems to be

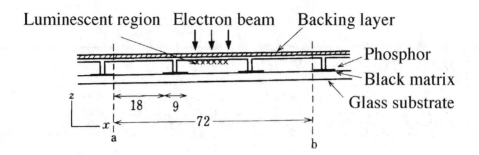

Figure 12 Cross-section of the phosphor screen of a color picture tube. Calculations were carried out for the region between a and b. (From Urabe, K., *Jpn. J. Appl. Phys.*, 19, 885, 1980. With permission.)

the first application of a Monte Carlo method to powder screens, although such methods had been used previously to evaluate light scattering in clouds,[10] oceans,[11] and emulsion films.[12]

Here, the phosphor screen model shown in Figure 12 provides an example. Various quantities, from excitation to final light output, that are suitable for analysis by the Monte Carlo technique are as follows:

- Incident point of an excitation beam at the screen surface, that is, selection of the excitation beam.
- Direction of a beam scattered at a screen component (backing layer, particle, black matrix, or glass substrate).
- Energy absorbed by a small part of a particle: its location in the particle.
- Energy absorbed at the surfaces of glass substrate, black matrix, and backing layer.
- Direction, polarization, and intensity of a light ray emitted from a small portion of a particle.
- Fraction of the emission that is absorbed in the particle.
- Direction and intensity of a ray scattered at the inner surface of the particle: location of scattering site.
- Absorption at various components on which the emitted ray falls.
- Intensity and angle of light scattered at the surface of each component.

As the result of the calculation, the intensities and directions of light emitted outward from the screen can be obtained as functions of position on the screen surface.

Two types of screen models are considered here. One, the particle layer model, uses a screen composed of individual particles and other entities. This model is suitable for thin screens and screens with complicated structures. The other, the continuum model, uses as a screen a three-dimensional continuum with absorption and scattering parameters characterizing constituent particles. This is a statistical approach and is used to model thick layers.

3.3.2.1 Particle layer model

Model of a phosphor screen. Urabe[7] used a phosphor screen model like that shown in Figure 12. The screen consists of spherical phosphor particles arranged in stripe patterns. Two types of particle arrangements (Figure 13(a) and (b)) were studied: (a) a regular arrangement of particles of the same size, and (b) an irregular arrangement of particles of three sizes. These arrangements are repeated in the y-direction, and the unit of length is the diameter of the sphere in Figure 13(a).

Chapter three: Optical properties of powder layers

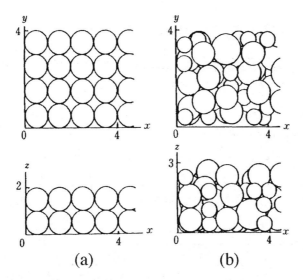

Figure 13 Packing of particles in the screen: (a) Regular packing of spheres with the same diameter: 1. (b) Irregular packing of spheres with three different diameters: 0.7, 1.0, and 1.3. Screen weights are the same in (a) and (b). (From Urabe, K., *Jpn. J. Appl. Phys.*, 19, 885, 1980. With permission.)

Principles of calculation.   Rays from excited phosphor particles are traced by geometrical optics until they are absorbed in the screen or escape from the screen. It is assumed that when a ray hits a surface, only one event—reflection, refraction, or absorption—occurs. The intensity of a ray remains constant unless it is absorbed. The flow chart of the calculation is shown in Figure 14. Other assumptions simplifying the calculation are as follows:

- The electron beam passing through the backing layer irradiates the particles, the substrate glass, or the black matrix. The beam power is completely absorbed there; scattering and penetration of the beam are not considered.
- The luminescence intensity of each particle is proportional to the power it absorbs. Emitted light is unpolarized and is normal to the surface of the excited particle. The angular distribution of its intensity is uniform.
- Reflection of light at the surfaces of the backing layer, glass substrate, and black matrix is specular, and reflectivities are independent of the incident angle of the light. Absorption in the glass is not considered.
- A constant fraction of the light incident on a particle is absorbed at the surface but is not absorbed inside the particle. The reflection and refraction of the residual light at the particle surface are given by Fresnel's law. Four internal reflections are taken into account in the calculation.

Application of the Monte Carlo method.   Random numbers R_i ($i = 1, 2, \ldots$) uniformly distributed between 0 and 1 determine the following items:

1. Direction of rays emitted with a uniform angular distribution: the direction (θ,ϕ) of a ray is calculated from two random numbers R_1 and R_2 as:

$$\theta = \cos^{-1}(1 - 2R_1) \tag{100}$$

$$\phi = 2\pi R_2 \tag{101}$$

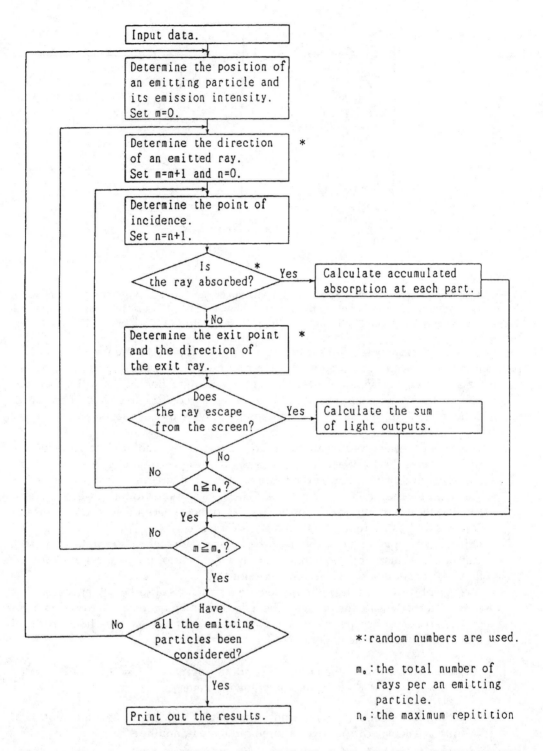

Figure 14 Flow chart of the calculations. (From Urabe, K., *Jpn. J. Appl. Phys.*, 19, 885, 1980. With permission.)

Chapter three: Optical properties of powder layers

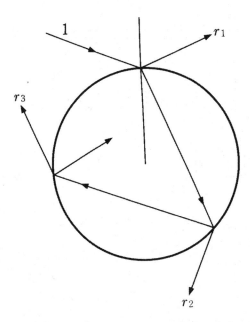

Figure 15 Light reflection and refraction at a particle surface. (From Urabe, K., *Jpn. J. Appl. Phys.*, 19, 885, 1980. With permission.)

2. Selection of events: which process, i.e., absorption, reflection, or refraction, occurs when a ray falls on a surface is also determined by random numbers.

A ray is completely absorbed at the incident point if a random number R_3 is smaller than the absorption ratio k. For the backing layer and black matrix, k = (1 − reflectivity) and a nonabsorbed ray is reflected. At the glass substrate, a ray is reflected when R_4 ð reflectivity; otherwise, it is transmitted. A ray that hits a particle surface and is not absorbed at the incident point suffers an i number (i = 1, ..., m; here, m is taken to be 6) of reflections and/or refractions at the outer or inner surface before leaving the particle (Figure 15). The intensity r_i of this departing ray, with initial intensity 1, is calculated by Fresnel's formula and is interpreted as the escape probability. The point on the surface and the direction of the escaping beam are obtained geometrically. The number i is determined by a random number R_5 as:

$$\sum_{j=0}^{i-1} r_j/R < R_5 \le \sum_{j=0}^{i} r_j/R \tag{102}$$

where

$$r_0 = 0, \quad R = r_1 + r_2 + ... + r_m$$

Results of calculation. The luminescent light output from the screen and the quantities absorbed by various components are obtained in relation to the position of the emitting particle. Figures 16 and 17 show the fractions of light emitted to the outside or absorbed by various screen components when the phosphor particles are packed regularly.

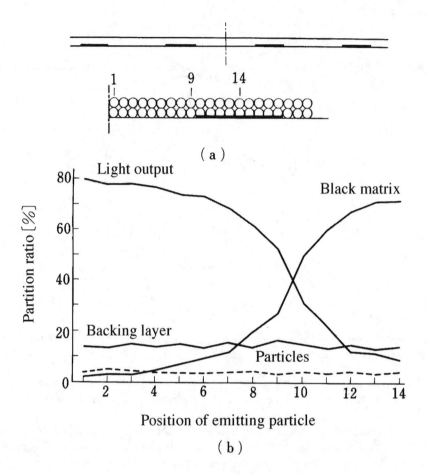

Figure 16 (a) Crosssection of fully coated phosphor screen. (b) Partition ratios of luminescent emission from a phosphor particle in model (a). (From Urabe, K., *Jpn. J. Appl. Phys.*, 19, 885, 1980. With permission.)

Any required information can be obtained by these types of calculations using the appropriate parametric values. Figure 18, for example, shows the partitioning of the total emission and the relative brightness as functions of the width of the emitting region, and Table 6 lists the partition ratios of the total emission for regular and irregular packing arrangements.

3.3.2.2 Continuum model

Soules and Klatt[8] studied the optical properties of powder coatings with spherical particles and of those with polyhedral particles. The former case is described here to show how the Monte Carlo method is applied. For a sphere of diameter d and a relative index of refraction $n = n_r - n_i i$, the scattering intensities for parallel and perpendicular polarization, $I_p(\theta)$ and $I_s(\theta)$, are calculated by the Mie theory for 181 values of the angle θ. The scattering cross-section efficiency Q_{sca} and extinction cross-section efficiency Q_{ext} are also calculated. The distance to the first scattering event is chosen as:

$$s = -(Q_{sca}/N)\ln(R_1) \tag{103}$$

where N is the average number of particle layers and R_i is a random number between 0 and 1. The angles θ and ω of the scattered ray are determined by two other random numbers:

Chapter three: Optical properties of powder layers

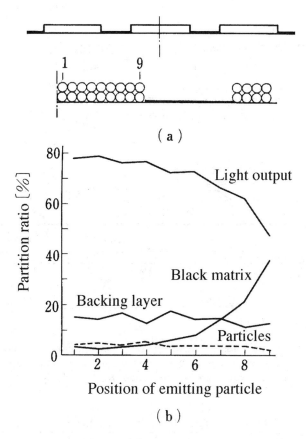

Figure 17 (a) Cross-section of a partially coated phosphor screen. (b) Partition ratios of luminescent emission from a phosphor particle in model (a). (From Urabe, K., *Jpn. J. Appl. Phys.*, 19, 885, 1980. With permission.)

$$\omega = 2\pi(R_2) \quad (104)$$

$$R_3 = \sum_{0}^{\theta}\left(f_p I_p(\theta) + f_s I_s(\theta)\right) \Big/ \sum_{0}^{180}\left(f_p I_p(\theta) + f_s I_s(\theta)\right) \quad (105)$$

The intensity of the scattered ray is attenuated by Q_{sca}/Q_{ext}, and the polarization of the scattered radiation in the plane of scattering is determined by the ratio I_p/I_s times the fraction of the incident polarization intensities f_p and f_s. The distance to the next scattering event is determined by another random number, and the ray is followed until it leaves the boundary of the coating or is attenuated to 0.0001 of its original intensity. The light absorption of various phosphor coatings is obtained in this way. Soules and Klatt[8] proposed a relation between the absorption coefficient of ultraviolet light and the optimum particle size of phosphors for a fluorescent lamp; they also used the Monte Carlo technique for particles with size distributions.

Busselt and Raue[9] used a continuum model for a uniform phosphor screen of a cathode-ray tube. The electron beam generates photons in the powder layer near the Al film and electron scattering in the powder is ignored. A generated photon is deemed to be a particle and to make a random walk (i.e., to move in steps whose lengths and

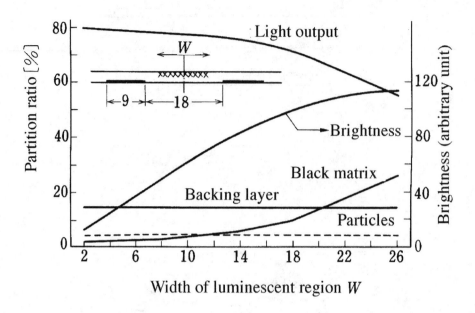

Figure 18 Dependence of partition ratios of total luminescent emission and of brightness on the width of the luminescence region. The unit of the abscissa is particle diameter and the packing of the phosphor screen is as in Figure 16. (From Urabe, K., *Jpn. J. Appl. Phys.*, 19, 885, 1980. With permission.)

Table 6 Partition Ratios for Two Packing Modes

Packing of particles	Partition ratio (%)			
	Light output	Backing layer	Black matrix	Particles
(a)	69.9	14.5	11.4	3.8
(b)	70.1	13.2	13.7	3.0

From Urabe, K., *Jpn. J. Appl. Phys.*, 19, 885, 1980. With permission.

directions are random). A Monte Carlo technique is used in the ray tracing to determine an absorption length, x, according to the probability distribution $p(x) = A^{-1}\exp(-x/A)$, where A is the absorption coefficient. A scattering length, y, is similarly determined according to the probability distribution $p(y) = S^{-1}\exp(-y/S)$, where S is the scattering coefficient. The absorption and the total scattering lengths (i.e., the sum of all previous scattering lengths) are compared after each scattering event, and the photon is assumed to be absorbed if the total scattering length is greater than or equal to the absorption length. A photon hitting the Al film is reflected or absorbed according to the reflectance of the Al film. A new scattering length and a new direction of the photon are chosen subsequent to each scattering event, and the angular distribution is assumed to be isotropic. This procedure is repeated until the photon is absorbed or escapes from the phosphor layer. An escaped photon is classified according to its position and angle of emergence. The screen efficiency, the angular distribution of photons leaving the screen, and the line spread function are obtained from these calculations. On the basis of these calculations and the results of measuring brightness and resolution of the screen experimentally, Bussel and Raue proposed a relation between the optimum screen weight and the brightness and a relation between the maximum layer thickness and the resolution.

3.3.3 Further problems

The following problems are left to be studied using Monte Carlo methods:

1. Calculation of scattering and absorption of excitation beams in a phosphor particle.
2. Calculation of the position on a spherical particle surface from which luminescent light is emitted and of the direction and intensity of the emitted light.

Several parameters need to be introduced to represent shapes other than spheres and to represent irregular or rough surface conditions.

Some additional work has been carried in this area. A Monte Carlo based simulation of the effects of coating on phosphor particles showed, for example, that thick coatings could increase the absorptivity of the phosphors. Comparison of this modeling to reflectance data on Eu^{3+}-activated yttria allowed the extraction of an absorption coefficient and allowed the simulations to make contact with a simplified Kebulka–Munk relationship.[13]

References

1. Chu, C.M. and Churchill, S.W., *J. Phys. Chem.*, 59, 855, 1955.
2. Emslie, A.G. and Aronson, J.R., *Appl. Opt.*, 12, 2563, 1973.
3. Meador, W.E. and Weaver, W.R., *Appl. Opt.*, 15, 3155, 1976.
4. Egan, W.G. and Hilgeman, T., *Appl. Opt.*, 17, 245, 1978.
5. Mudgett, P.S. and Richard, L.W., *Appl. Opt.*, 10, 1485, 1971.
6. Soules, T.F., *Electrochem. Soc. Extended Abstracts*, 74-1, 311, 1974.
7. Urabe, K., *Jpn. J. Appl. Phys.*, 19, 885, 1980.
8. Soules, T.F. and Klatt, W.A., *J. Illum. Eng. Soc.*, 17, 92, 1988.
9. Busselt, W. and Raue, R., *J. Electrochem. Soc.*, 135, 764, 1988.
10. Plass, G.N. and Kattawar, G.W., *Appl. Opt.*, 7, 415, 1968.
11. Gordon, H.R. and Brown, O.B., *Appl. Opt.*, 12, 1549, 1973.
12. DePalma, J.J. and Gasper, J., *Photgr. Sci. Eng.*, 16, 181, 1972.
13. Ling, M. and Soules, T.F., *Electrochem. Soc. Extended Abstracts*, 86, 1038, 1988.

chapter four — sections one–seven

Color vision

Kohei Narisada and Sueko Kanaya

Contents

4.1 Color vision and the eye 116
    4.1.1 Retina 116
    4.1.2 Spectral sensitivity of the eye 117
4.2 Light and color 118
    4.2.1 Psychophysical colors 118
    4.2.2 Perceived colors 119
4.3 Models of color vision 119
4.4 Specification of colors and the color systems 121
    4.4.1 Specification of the psychophysical colors 121
        4.4.1.1 CIE colorimetric system 122
        4.4.1.2 CIE UCS colorimetric system 125
        4.4.1.3 ULCS colorimetric system 126
    4.4.2 Specification of perceived colors 128
        4.4.2.1 Munsell color system 128
        4.4.2.2 NCS 128
4.5 The color of light and color temperature 128
    4.5.1 Chromatic adaptation and colors of light sources 128
    4.5.2 Color temperature 129
    4.5.3 Mired (micro-reciprocal degree) 129
4.6 Color rendering 130
    4.6.1 Methods of measurement 130
    4.6.2 Color rendering index 131
        4.6.2.1 General color rendering index (R_a) 131
        4.6.2.2 Special color rendering indices 131
    4.6.3 General color rendering index and perceived colors 131
    4.6.4 Color appearance of light sources and perceived colors 132
    4.6.5 Color rendering and brightness 132
4.7 Other chromatic phenomena 132
    4.7.1 Purkinje phenomenon 132
    4.7.2 Metamerism 133
    4.7.3 Bezold-Brucke effect 133
    4.7.4 Helmholz-Kouroshe effect 133
References 133

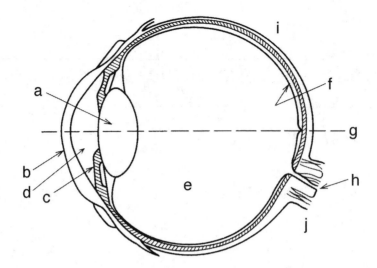

Figure 1 Schematic of the horizontal cross-section of the human eye. a: lens, b: cornea, c: iris, d: aqueous humor, e: vitreous humor, f: retina, g: visual axis, h: optic nerve, i: ear side, j: nose side.

Human eyes have two fundamental visual functions: to perceive brightness and to detect color.[1] In this chapter, the physiological mechanism of color vision and the basis for the specification of colors will be outlined. Some associated psychological phenomena will also be briefly described.

4.1 Color vision and the eye

The organ of human vision consists of the eyeball, the visual center of the brain, and the optic nerves connecting the two. As schematically shown in Figure 1, the eyeball is an optical system consisting of the cornea, aqueous humor, iris, crystalline lens, vitreous humor, and the retina. By means of refraction, the cornea and the crystalline lens focus an upside-down optical image of the outer world within the field of vision on the retina. Optical signals corresponding to the optical image on the retina are transformed into neural signals and sent through the optic nerves to the visual center of the brain.

4.1.1 Retina

The retina is a light-sensitive organ made up of an extremely large number of optic receptors connected to each other by a sophisticated neural network. There are two types of the receptors having different shapes and physiological functions. The light-sensitive portion of the first receptor has a relatively short and thick shape (Figure 2(a)), while the other type has a slender and thin shape (Figure 2(b)). From these shapes, the former are called cones and the latter rods. There are three kinds of cones, each of which is sensitive to the optical radiation in different wavelength ranges, i.e., red (R), green (G), and blue (B) ranges. These three kinds of cones are distributed over the entire retina, except in the fovea where almost no blue- sensitive cones exist with an angular extent of about 2°. The fovea, on the other hand, has rods that are all the same and are not sensitive to different colors; consequently, rods have no function in perceiving colors.

The sensitivity of the cones to optical radiation is relatively low in comparison with that of the rods. The cones are fully active when there is a field of view with a

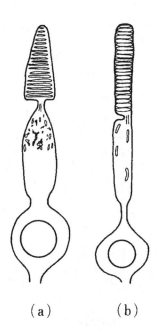

Figure 2 Sketch of two types of visual receptor cells: (a) cone receptor; (b) rod receptor.

luminance higher than about 1 cd m^{-2}. The rods, on the other hand, have a higher sensitivity and are fully active in dark environments, having luminance of less than ~0.01 cd m^{-2}. The visual condition in bright environments with luminances of about 1 cd m^{-2}, under which only the cones are active, is called *photopic vision*. The opposite condition in dark environments with luminance less than 0.01 cd m^{-2}, where only the rods are active, is called *scotopic vision*. Intermediate conditions, where the luminance range in the field of view is between 0.01 and 1 cd m^{-2}, and where the cones as well as the rods are partially active, is called *mesopic vision*. The roles of the two types of receptors—the cones and the rods—vary gradually as the luminance level in the field of view changes.

4.1.2 Spectral sensitivity of the eye

The spectral sensitivity of the eye is not solely determined by that of the retina. The spectral transmission characteristics of the optical components of the eye (i.e., cornea, crystalline lens, and humors) also play some role. As a whole, the eye is sensitive to optical radiation with a wavelength between 380 and 760 nm when it is incident on the eyeball through the cornea. The sensitivity over this wavelength range, however, is not constant and has specific spectral characteristics. This spectral dependence is called the spectral luminous efficiency function of the eye.

The eye has two different spectral luminous efficiency functions corresponding to the cones or the rods. These two sensitivity curves are shown in Figure 3 as a function of the wavelength, and are similar in shape but have different peak wavelengths. Strictly speaking, the eye of each individual has a different spectral luminous efficiency function depending on such factors as age, etc. To deal with the light quantitatively, however, the international standard visual spectral efficiency curves were established by the CIE (Commission Internationale de l'Eclairage) and the ISO (International Standardization Organization) and are used for photopic vision $V(\lambda)$ and scotopic vision $V'(\lambda)$, respectively,[1] as

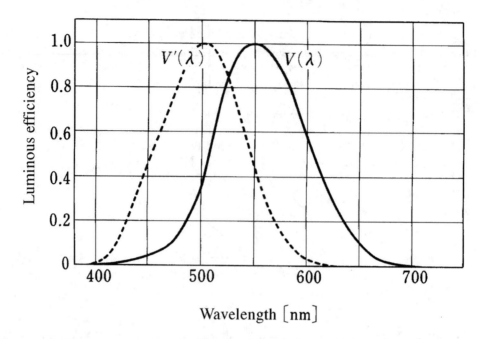

Figure 3 Standard visual spectral efficiency curves, photopic vision V(λ), and scotopic vision V'(λ). From CIE, *The Basis of Physical Photometry*, CIE Publication No. 18.2, 1983. With permission.)

shown in Figure 3. The two standard visual spectral efficiency curves in the figure are given relative to the maximum value of 1 at the wavelength of peak sensitivity. The absolute maximum value of the luminous efficiency at the peak wavelength of 555 nm is 683 lm W^{-1} for photopic vision V(λ) and 1700 lm W^{-1} at 507 nm for scotopic vision V'(λ). The mesopic vision is thought to have intermediate spectral sensitivity characteristics. Presently, no definite standard has been established for mesopic vision, since, as mentioned previously, it varies gradually as the luminance changes.

Photometric quantities such as the luminous intensity (cd), the luminous flux (lm), the illuminance (lx), and the luminance (cd m^{-2}), etc. are derived after integrating the energy of the optical radiation at different wavelengths over the standard spectral efficiency function for photopic vision.

4.2 Light and color

As previously stated, the eye is sensitive to optical radiation with a wavelength range between 380 and 760 nm. Optical radiation within this wavelength range is called visible light. The eye perceives different colors for different wavelengths. Colors with a single wavelength or those with a narrow range of different wavelengths are called pure colors or monochromatic. The visual sensation for perceived colors is a composite of the sensations for the different pure colors that make up the color. Colors are divided into two categories: phychophysical and perceived colors.

4.2.1 Psychophysical colors

The psychophysical colors are those that can be treated and measured in terms of the chromatic response of the eye to pure colors of different wavelengths. The response is derived by psychophysical color matching experiments to be described below.

4.2.2 Perceived colors

The perceived colors are those subjectively perceived by the eye. These colors cannot be measured quantitatively. To specify a perceived color, visual comparisons are made to find a match to the color among a series of specific color samples arranged visually in a graded scale. The perceived colors are divided further into three subcategories: object, luminous, and aperture colors.

The object colors are those belonging to the object. They are classified by the principal hue, the lightness, and the chroma of the colors. The light source or luminous colors are those of produced by light emitting-objects. Finally, the aperture colors are those of a lighted surface as seen through a small aperture. The human eye perceives the same colors differently according to various environmental, physiological, and psychological conditions.

4.3 Models of color vision

It is thought that human beings have had the same color vision for thousands of years. It is found that primitive pictures of natural objects discovered in a number of archaeological ruins, some tens of thousand years old, were painted in colors somewhat similar to those observed at present. In the earlier stages of human history, such as the periods of ancient Egypt or Sumer, people already were interested in colors and color perception. Since then, various ideas on color and color perception have been advanced; among these are works by Leonardo da Vinci[2] in 1651 and Newton[3] in 1704.

The first theory directly connected to modern theories on color perception or color vision was presented by Young in 1802.[4] This theory was extended by Helmholtz[5] in 1852. Today, their theories have been combined and are jointly called the Young-Helmholtz's trichromatic theory. This theory gave an explanation for the perception of mixed colors. It could not explain, however, some phenomena concerning the perception of colors and chromatic functions of defective color vision. One of the problems that could not be explained was why yellow appears as one of the primary colors among various perceptible colors, and why the four primary colors are made up of two sets of opposite colors, red-green and blue-yellow. For this reason, a number of ideas opposing the trichromatic theory were proposed. These opposite colors are called opponent colors. Hering[6] addressed the phenomenon concerning the opponent colors in 1875. He pointed out that colors were perceived as a mixture of each of the opponent colors of these two sets. On the basis of this idea, a model for color vision called the opponent color theory was proposed.

The trichromatic theory and the opponent color theory contain sharply contrasting ideas on this matter. Later in 1896, Müller et al.[7] assumed that the three kinds of receptors, each responding to one of the three primary colors, act in the way the trichromatic theory predicts. In the neural network of the retina, however, three different neural signals are activated by the three types of optic receptors, but are processed and transformed as two sets of opponent color signals, consistent with the opponent color theory. This was the beginning of the zone theory.

From the model by Müller, Walraven, and Bouman[8] proposed in 1966, a schematic model explaining the zone theory for color vision. Later, in 1980, the model was slightly modified by Ikeda,[9] as shown in Figure 4. Though the theory has not yet been proven conclusively, in the following with reference to the figure, the likely process for color vision is explained in a simplified way (for further details, the original publications should be referred to). In Figure 4, boxes represent hypothetical components of the neural system, which can add or subtract input signals and send their outcomes to the next stages. Lines

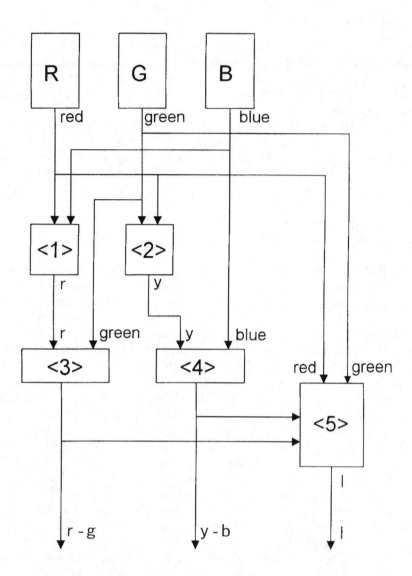

Figure 4 A model for color vision. (From Ikeda, M., *Fundamentals of Colour Engineering*, Asakura Bookstore, 1980, 244-245. With permission.)

with arrows show the direction of flow of the neural signals. Three boxes, R, G, and B on top of the diagram, represent three types of cones, each sensitive to one of the three primary colors—red, green, and blue, respectively. If the cones, as a group, are stimulated by a color, each of the three cones generates a neural signal proportional to the primary colors contained in the color stimulus.

Among the three neural signals, the red and green signals are sent to <5> and added to each other. The signal I is sent to the brain. Precisely speaking, signals from rods under mesopic or scotopic conditions are also sent to <5> to register as the brightness signal I. At this stage, the blue signal from cone B is not sent to this channel, since it contains only a color signal, but no brightness signal.

The signals red and blue from R and B, respectively, are added in <1> and a signal r, corresponding to real red (a temporary expression), is created. Similarly, in <2>, the signals

Chapter four: Color vision

green and blue from G and B, respectively, are mixed and a signal y, corresponding to yellow, is formed. The intensity of the two signals green and r from <1> are compared in <3> and only the stronger signal is sent to the brain. Such a comparison of the two signals produces the so-called opponent signal expressed as r-g. In this case, either the r (real red) or the green signal only is sent to the brain.

In the same way, the intensities of the two signals y (yellow) from <2> and blue from B are compared to form the opponent signal y-b. Again, the stronger of the two will be sent to the brain. Consequently, the eye sends to the brain a brightness signal and two color signals composed of one of the components of the opponent colors.

It is interesting to note (see the figure) that the two opponent signals r-g and y-b flow into <5> and contribute to the brightness channel. Since no opponent signal concerning color will be sent to the brain when the eye sees a white object, the brightness detected is influenced by the color of the object.

In the experimental process through which the standard luminous efficiency functions shown in Figure 3 were derived, however, the contribution of chromatic signals to the brightness has not been incorporated. For this reason, even if the same luminance value is obtained through the function for a colored and a non-colored object, the perceived brightness of the two objects is not necessarily the same. The matter is still the subject of active discussion among color scientists.[10]

4.4 Specification of colors and the color systems

Color is quantitatively specified by color systems. Color systems are divided into two categories. The first specifies psychophysical colors while the second deals with perceived colors.

Psychophysical colors are specified by the principle of additive mixture of stimuli of the three primary colors. As mentioned previously, colors are composed of primary colors, and their mixing ratios determine the colors. This means that a color can be specified with a mixing ratio of two of the three primary colors; since the sum of the primary color must always be 100%, the specification of the mixing ratio of two automatically determines the remaining primary color. A representative system is the CIE color system, which specifies colors by two coordinates (equivalent to the designation of the mixing ratio of two primary colors) on a chromaticity diagram. This is discussed 4.4.1.1. The position of a color in the diagram is called the chromaticity point of the color.

Systems that specify the perceived colors, on the other hand, classify and scale many colors systematically according to their sensual attributes, such as hue, lightness, and chroma. Two color systems, the Munsell Color System and the NCS (Natural Color System), are in current use.

4.4.1 Specification of the psychophysical colors

Psychophysical colors are specified by three systems: the CIE, the CIE UCS (Uniform Chromaticity Scale), and the CIE ULCS (Uniform Lightness Chromaticness Scale) colorimetric systems. Each system specifies colors using the coordinates of a chromaticity diagram to be described later. The UCS colorimetric system evolved from the CIE system. In the UCS, the distance between two chromaticity points on the chromaticity diagram corresponds to the sensation of perceived color difference. The color difference on the UCS colorimetric coordinates, however, does not correspond to the difference in perceived colors if the lightness of the two colors differs. The ULCS colorimetric system was then introduced to compensate for the difference in the lightness of the colors. A number of programs to be used in personal computers have been developed for calculations associated with the use of these color systems.

4.4.1.1 CIE colorimetric system

The CIE colorimetric system was established by the CIE in 1931. This system consists of the RGB and the XYZ colorimetric systems. The XYZ system, which will also be explained below, was laid down as an extension of the RGB system for practical applications.

The RGB system was derived from results of psychophysical experiments. In the experiments, the observers viewed a circular field with an angular diameter of 2°. (The angular diameter is the angle of arc subtended by the circular field at the eye.) The circular field consisted of two identical half circles adjacently located on the right and left. The color of the two half circles was independently variable. One of the two half circles was used as the reference field and another was used as the test field.

Colors of the reference field were called the reference colors and those of the test field were called the test colors. The reference colors were expressed by monochromatic light of the same intensity at various wavelengths over the entire visible range. The test colors, on the other hand, were composed with a mixture of the three primary colors, red (700 nm), green (546.1 nm), and blue (453.8 nm). The numbers in parentheses show the wavelengths of the respective primary colors.

By varying the mixing ratio of the three primary colors, the observers varied the colors of the test field. In this way, the color of the test field was made to match that of the adjacent reference field. During the observations, it was found that in some wavelength ranges mixtures of the three primary colors could not match the reference colors. In these wavelength ranges, matches were established if an amount of one of the three primary colors was added to the monochromatic reference colors.

This implies that matches can be established by subtracting one of the primary colors from the mixtures. In other words, there are some wavelength ranges where the stimulus of the primary colors is negative. In this way, the mixture ratio of the primary colors to match all the spectral colors over the entire visible range were obtained.

It was assumed, when a match was established, that the reciprocals of the energy ratio of each primary color of the test field corresponded to the relative strength of the stimuli of the respective primary colors at the wavelength of the reference with which the color was matched. Based on the above assumption, three spectral distribution curves of the relative strength of the stimulus for each of the three primary colors (red, green, and blue) over the entire visible range were obtained. The curves are called the spectral tristimulus values or the color matching coefficients. They are $r(\lambda)$, $g(\lambda)$, and $b(\lambda)$, respectively. λ in the parentheses is the wavelength.

Since, the sum of the three tristimulus values at each wavelength is always 100%, the mixture ratio of the three primary colors can be determined by any two of the three tristimulus values. The RGB colormetric system is based on this and all colors are indicated on the $r(\lambda)$ and $g(\lambda)$ coordinates.

To overcome difficulties associated with the negative stimulus of the primary colors, based on the above mentioned color matching experiments, three imaginary reference color stimuli [X], [Y], and [Z] were introduced. By employing the imaginary reference color stimuli, the original tristimulus values were converted mathematically into positive values and all colors could be composed by mixing (not subtracting) these three stimuli. This is the basis of the XYZ colorimetric system. Y has been chosen to correspond with the lightness stimulus. Based on the similar idea with that of the RGB colorimetric system, all colors are indicated by these coordinates.

More recently, similar color matching observations have been conducted for a larger circular field with an angular diameter of 10°. From the results obtained, a colorimetric

Chapter four: Color vision

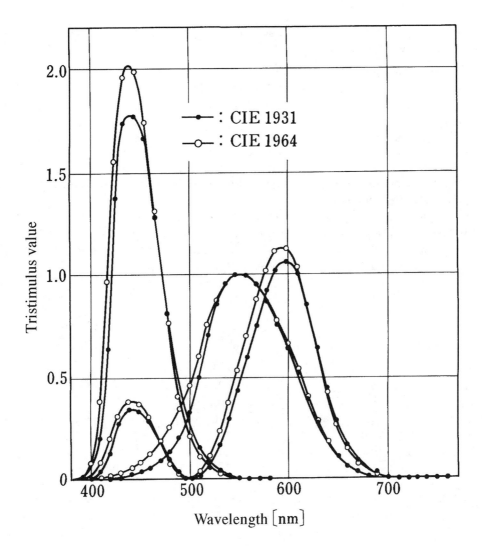

Figure 5 Curves of spectral tristimulus values. ○: CIE 1964 supplementary standard colorimetric observer; ●: CIE 1931 standard colorimetric observer. (From CIE, *Testing of Supplementary Systems of Photometry*, TC1-21, in preparation. With permission.)

system was established by the CIE in 1964. The results of these experiments are shown in Figure 5. In the figure, the curves marked with black dots show the spectral tristimulus values for the circular field of 2° established in 1931, which can be used for a field smaller than about 4°; and the curves marked with white circles are those for 10° established in 1964, which can be used for a field greater than about 4°. Each curve is used according to the size of the object field in which the color is to be specified.

Specification of light source colors. Test light source colors are specified below: the tristimulus values (X, Y, and Z) for a test light source, which has a spectral energy distribution P(λ), are calculated with the following formulae.

$$X = K \int_{380}^{780} P(\lambda)\bar{x}(\lambda)d\lambda \tag{1a}$$

$$Y = KP(\lambda)\bar{y}(\lambda)d\lambda \qquad (1b)$$

$$Z = K\int_{380}^{780} P(\lambda)\bar{z}(\lambda)d\lambda \qquad (1c)$$

where
$$K = \frac{1}{P(\lambda)Y(\lambda)d\lambda}$$

and $x(\lambda)$, $y(\lambda)$, and $z(\lambda)$ are the spectral stimulus values for 2°. These quantities are written as $x_{10}(\lambda)$, $y_{10}(\lambda)$, $z_{10}(\lambda)$ for 10.

The chromaticity coordinates of the color of the light sources x and y are calculated with the following formulae.

$$x = \frac{X}{X+Y+Z} \qquad (2a)$$

$$y = \frac{Y}{X+Y+Z} \qquad (2b)$$

The colors of light sources on the XYZ colorimetric system are specified with Y calculated with Eq. 1b and with x and y calculated with Eq. 2a, b.

Figure 6 shows the CIE chromaticity diagram in which the ISCC-NBS (Intersociety Color Council–National Bureau of Standard) color designation for every color[11] is indicated. The number beside the spectrum locus shows the wavelength of the monochromatic light. Points labeled Illuminant A and Illuminant C corresponding to color temperature of 2854K and 6774K, respectively, are shown in the central white region.

Specification of the nonluminous object colors. The tristimulus values (X, Y, and Z) of the object for which the spectral reflectance (or spectral transmittance) is $\rho(\lambda)$ and $\tau(\lambda)$ are given by:

$$X = \frac{1}{K}\int_{380}^{780} P(\lambda)\rho(\lambda)\bar{x}(\lambda)d\lambda \qquad (3a)$$

$$Y = \frac{1}{K}\int_{380}^{780} P(\lambda)\rho(\lambda)\bar{y}(\lambda)d\lambda \qquad (3b)$$

$$Z = \frac{1}{K}\int_{380}^{780} P(\lambda)\rho(\lambda)\bar{z}(\lambda)d\lambda \qquad (3c)$$

where
$$K = \int_{380}^{780} P(\lambda)\bar{y}(\lambda)d\lambda \qquad (3d)$$

$P(\lambda)$ is the spectral power distribution of the light source which illuminates the object, and $x(\lambda)$, $y(\lambda)$, and $z(\lambda)$ are the CIE spectral trichromatic stimuli for fields of 2° or 10°. The chromaticity coordinates of the color of the objects can then be calculated, as with the light sources, using Eq. 2.

Chapter four: Color vision

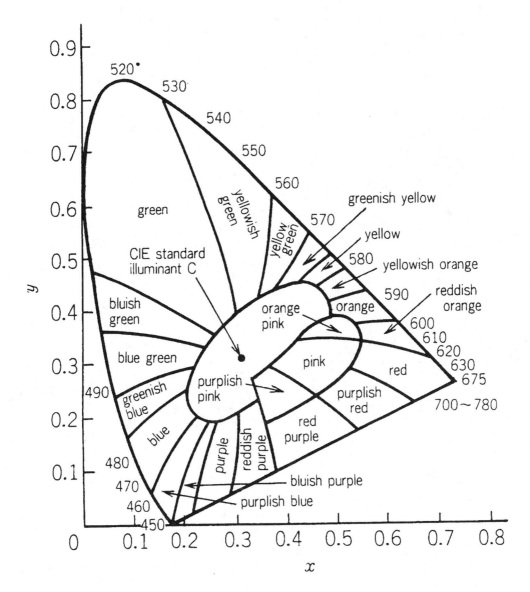

Figure 6 CIE Chromaticity diagram. (From Kelly, K.L., *J. Opt. Soc. Am.*, 33, 627, 1943. With permission.)

4.4.1.2 CIE UCS colorimetric system

The distance between two chromaticity points on the chromaticity diagram gives the color difference. The extent of the color difference on the CIE chromaticity diagram shown in Figure 6, however, does not correspond to the actual perceived color difference. This means that for two pairs of colors with the same distance between their chromatic points but in different areas of the diagram are not perceived to have the same color difference.

The UCS chromaticity diagram was developed to overcome this problem by modifying the chromaticity coordinates to make the distance between any two color points correspond to the perceived color difference. In 1960, based on the assumption that the standard deviation of the data on the matching observations corresponded to the color discrimination threshold, the CIE established a modified chromaticity diagram. This is the UCS diagram shown in Figure 7.

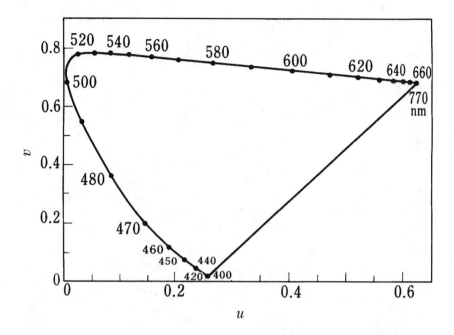

Figure 7 CIE UCS 1960 chromaticity diagram. (From CIE, *Testing of Supplementary Systems of Photometry*, TC1-21, in preparation. With permission.)

The chromaticity coordinates of the UCS chromaticity diagram (u, v) can be obtained by transformation of the x, y, and z coordinates for the XYZ chromaticity diagram into new coordinates, u and v, using the transformations in Eqs. 4a and 4b.

$$u = \frac{4x}{-2x + 12y + 3} \tag{4a}$$

$$v = \frac{6y}{-2x + 12y + 3} \tag{4a}$$

The coordinates u and v can also be obtained from X, Y, and Z from the XYZ chromaticity diagram using the following formulae.

$$u = \frac{4X}{X + 15Y + 3Z} \tag{5a}$$

$$v = \frac{6Y}{X + 15Y + 3Z} \tag{5b}$$

In 1976, the CIE established the u′, v′ color system as an improvement to the UCS system. The new coordinates u′ and v′ can be obtained simply by u′ = u and v′ = 1.5v.

4.4.1.3   ULCS colorimetric system

As explained above, the UCS colorimetric system was developed to make the distance between the two color points on the chromaticity diagram correspond to actual perceived

Chapter four: Color vision

color differences. The system is valid, however, only when the lightness of the colors is the same.

The ULCS colorimetric system was introduced to improve the UCS system. At present, the CIE recommends two ULCS colorimetric systems. They are the CIE 1976 L*u*v* Color Space and the CIE 1976 L*a*b* color space.

X, Y, and Z in the XYZ colorimetric system can be transformed into the coordinates in the L*u*v* color space using Eqs. 6a, b, and c, or into the L*a*b* color space using Eqs. 8a, b, c. The L*u*v* colorimetric system is plotted on an orthogonal coordinate system with axes given by:

$$L^* = 116 \left[\frac{Y}{Y_0} \right]^{1/3} - 16 \tag{6a}$$

$$u^* = 13 L^* (u' - u'_0) \tag{6b}$$

$$v^* = 13 L^* (v' - v'_0) \tag{6c}$$

where

$$u' = \frac{4X}{X + 15Y + 3Z} \tag{7a}$$

$$v' = \frac{9Y}{X + 15Y + 3Z} \tag{7b}$$

$$u'_0 = \frac{4X_0}{X_0 + 15Y_0 + 3Z_0} \tag{7c}$$

$$v'_0 = \frac{9Y_0}{X_0 + 15Y_0 + 3Z_0} \tag{7d}$$

X, Y, and Z refer to the trichromatic stimulus of the object, while X_0, Y_0, and Z_0 refer to the light source that illuminates the object. The value of Y_0 is normalized to 100. The value of L* corresponds to the psychometric lightness and is 10 times Munsell's value (V). As will be described below, the value V in the Munsell color system indicates the lightness of the colors. The values of u* and v* relate to the hue and the chroma in the Munsell color system, respectively, and are called the chromaticness index.

The L*a*b* colorimetric system gives the L*, a*, and b* values on the right-angle coordinates by the following formula.

$$L^* = 116 \left[\frac{Y}{Y_0} \right]^{1/3} - 16 \tag{8a}$$

$$a^* = 500 \left[\left(\frac{X}{X_0} \right)^{1/3} - \left(\frac{Y}{Y_0} \right)^{1/3} \right] \tag{8b}$$

$$b^* = 200\left[\left(\frac{Y}{Y_0}\right)^{1/3} - \left(\frac{Z}{Z_0}\right)^{1/3}\right] \quad (8c)$$

The value of L* corresponds to the psychometric lightness, and the values of a* and b* correspond to the hue and chroma in the Munsell color system, respectively. X, Y, and Z are the trichromatic stimulus of the object and X_0, Y_0, and Z_0 are the trichromatic stimulus of the light source that illuminates the object. The normal value of Y_0 is 100.

4.4.2 Specification of perceived colors

There are two systems that specify the perceived colors: the Munsell color system and the NCS (Natural Color System).

4.4.2.1 Munsell color system

The Munsell color system is based on the color chart in which the many colors are arranged systematically in equally perceived difference intervals for each of the three attributes of colors: hue (H), value (V), and chromaticness (C). The system was invented by Munsell in 1905 and formalized with some corrections in 1943 by the Optical Society of America. The colors are specified by a combination of quantities in the form HV/C (specifying the color properties quantitatively).

4.4.2.2 NCS

The NCS was proposed by Hard et al.[12] and initially specified as the Swedish National Standard by the Swedish Standards Institution (SIS). In this system, the colors are specified with the three attributes of whiteness (w), blackness (s), and chromaticness (c). The hues of the colors are indicated by the mixing ratios of the opponent colors red-green and yellow-blue. For example, a reddish-yellow that contains 70% yellow and 30% red is specified as Y_{30R}. The NCS system is based on the opponent color theory; thus, colors with a hue are specified by a mixing ratio of the three attributes whose sum is equal to 100. A pure color with a red hue, for example, is indicated as c = 100. Achromatic colors contain no chromaticness and are specified only by the ratio of the blackness (s) and the whiteness (w).

4.5 The color of light and color temperature

4.5.1 Chromatic adaptation and colors of light sources

Photometric measurements show that the color of natural daylight varies as the day progresses. Immediately after sunrise, it is very reddish. Under a very clear blue sky in the daytime, in the northern hemisphere, at the north side of buildings with no direct sunlight, daylight colors are bluish. Just before sunset, it again turns to a very reddish color. Nevertheless, human eyes are able to perceive colors in daylight hours with relative stability and accuracy, irrespective of the location. The reason is that the spectral sensitivity of the retina automatically compensates for changes in the color of the light illuminating an object. This physiological function of vision is called chromatic adaptation.

Chromatic adaptation is a function of vision that minimizes the influence of the color of illumination on the perception of colors of the object. Consequently, human eyes perceive the color of the light sources as being much more whitish than they actually are, depending on the chromatic environment or the state of the chromatic adaptation of the eye. This means that the chromaticity coordinates of colors of lights illuminating the object do not always correspond to the perceived colors. A number of methods to correct for the influence of the chromatic adaptation have been proposed by the CIE. It is important that chromatic adaptation does not compensate for the colors of light sources if lamps with different color temperatures are mixed in the field of view. Under such conditions, the lamps are seen in contrasting colors to one another; a bluish lamp is seen as more bluish, and a reddish lamp is seen as more reddish.

4.5.2 Color temperature

For the reasons mentioned above, it is necessary to define the color of the light in an objective way or by physical measurements. To define the color of light, the concept of the color temperature has been introduced. The color temperature is defined by the absolute temperature of a Planckian radiator (black body) that has the same color or the same colorimetric coordinates as that of the light source. Figure 8 shows the locus of the colorimetric coordinates of the color of the Planckian radiator in the CIE chromaticity diagram for a wide range of color temperatures. Further explanation of the locus will be given later. As seen in the figure, the color of the light sources changes from reddish to bluish color as the color temperature increases. When the chromaticity coordinates of a light source are not exactly on the locus of the Planckian radiator, the absolute temperature of the Planckian radiator closest to the light source is taken, and that temperature is called the correlated color temperature. A number of straight lines with designated temperatures are shown in the figure, each line being drawn perpendicular to the Planckian locus, and they show the closest line to the relevant temperature. Some examples of approximate figures of the color and the correlated color temperatures of natural and artificial light sources are shown in the table below.

Light source	Color temperature (K)
Blue sky	15000–20000
Cloudy sky	6500
Fluorescent lamps	
Daylight color	6500
Cool white color	4200
Halogen incandescent lamps	3000
Ordinary incandescent lamps	2800
High-pressure sodium lamps	2000
Candle flame	2000

4.5.3 Mired (micro-reciprocal degree)

The difference in the colors of lights with different color temperatures is indicated as the distance between their chromaticity points on the chromaticity diagram in Figure 8. As before, the difference in the perceived colors in different areas of the diagram is not the same, even if the distance on the chromaticity diagram is the same. To make the difference closer to the perceived color difference, a value of mired (micro-reciprocal degree) is used. Mired is defined as one million (10^6) times the reciprocal of the color temperature of the light source. If the difference in mired is the same, the difference in the perceived colors between the two light sources is roughly the same.

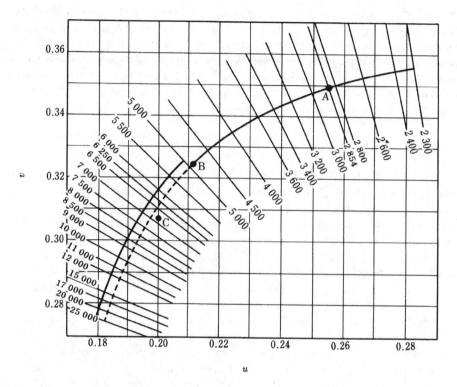

Figure 8 Locus of the color of the Planckian radiator in CIE 1964 chromaticity diagram for a wide range of color temperatures. (From CIE, *Testing of Supplementary Systems of Photometry*, TC1-21, in preparation. With permission.)

4.6 Color rendering

4.6.1 Methods of measurement

Color rendering is the property of the light source that changes the colors of the object illuminated by that light source. To deal with the color rendering, the following two methods are normally employed.

The first method is psychophysical. The basic approach of this method is to examine how the color of an object illuminated by a sample light source compares with the color observed under illumination with a reference light source having an ideal color rendering.

The second method, on the other hand, is psychological. The approach of this method is to examine by visual observation whether the color of an object under a sample light source is preferred by the observer or not. To conduct these observations, a careful selection of the colors to be examined is essential. Since this method is subjective, no definite method to determine this kind of color rendering has been developed. However, in some cases, the results derived by this method are practically more relevant and important than the results obtained through the psychophysical method.

To examine the color rendering properties of a sample light source based on the psychophysical method, colors of a number of objects illuminated by a sample light source are compared psychophysically with the colors obtained when the objects are illuminated by a reference light source. For this purpose, the chromaticity points of object colors are calculated with the CIE colorimetric system already described. Before the chromaticity points can be calculated for the reference light source, the spectral energy distributions and the color temperature of the reference light must be specified.

Chapter four: Color vision 131

The CIE specifies two series of Reference Light Sources to be used for these calculations. The first is a Planckian radiator and the second is the CIE "daylight" standard. To examine the color rendering of a sample light source, it is recommended that a reference light source with a color temperature lying within a range of 5 mireds of the correlated color temperature of the sample light source be used. For color temperatures lower than 5000K, it is recommended that the Planckian radiator be used. As a standard for temperatures higher than 5000K, on the other hand, the CIE daylight standard is recommended.

The two curved lines in Figure 8 show the loci of the chromaticity points of the Planckian radiator and the CIE daylight reference light sources, respectively. The color temperature is taken as a parameter. The dotted curve shows the overlap zone of the locus of the Planckian radiator with that of the CIE "daylight" standard.

4.6.2 Color rendering index

4.6.2.1 General color rendering index (R_a)

The color rendering index (R_a) indicates the extent of the color rendering properties of a light source. The basis of the calculation of the general color rendering index (R_a) of a sample light source is the use of color differences. This difference is the distance between the chromaticity points of the sample and the reference light source, and is obtained for each of eight selected object colors. R_a is calculated by taking the average of these differences. If no color difference is found for all eight object colors, then the general color rendering index (R_a) is the maximum, 100; it is observed that the larger the averaged difference is, the lower the R_a figure. To use these estimates, however, care must be taken as to the limits of applicability and the meanings of the indices. Careful interpretation of the indices is always necessary. Some of the pitfalls encountered are discussed below.

4.6.2.2 Special color rendering indices

Special color rendering indices are calculated for one of six test colors specified by the CIE. The special color rendering index of a sample light source for one of the test colors shows the closeness of that test color to that under a reference source having the same color temperature.

The special color rendering indices of a sample light source are calculated in a similar way to the general color rendering indexes. However, the special color rendering index is calculated individually for each of the following six selected colors. The six colors are selected to represent the colors of normal objects. They are red, yellow, green, blue, each with a high chroma, and the colors of a Caucasian complexion and a green leaf. The indices are expressed according to the selected colors as R_9, R_{10}, R_{11}, R_{12}, R_{13}, and R_{14}. In the Japanese standard, the color of an Oriental complexion is added as R_{15}.

4.6.3 General color rendering index and perceived colors

To use the general color rendering index R_a, some caution is necessary, since there is some limitation of its applicability. Some of these limitations are summarized below.

1. No valid comparisons can be made using only the general color rendering index R_a between lights with different color temperatures. A difference in R_a is valid only for light sources with similar color temperatures. If the color temperatures of the two light sources to be compared are considerably different, then the difference has no actual meaning. This is because the perceived colors under reference light sources vary for different color temperatures.

2. Judgment of the perceived colors cannot always be made between light sources with low R_a values. As described, the R_a value is calculated using an average of the color differences for the eight specified test colors. As can be easily seen, this average can be obtained with many different combinations of the color differences.

   This implies that the color rendering properties can be considerably different, even if the R_a value is the same. For example, one lamp may have color rendering properties that cause a color difference for all eight test colors nearly to the same extent. Another lamp, on the other hand, may have other color rendering properties that cause considerable color differences for some of the test colors, whereas there is almost no color difference for the remaining test colors. If the latter lamp is used in an application where only some colors are to be perceived correctly and the lamp causes no big color difference in these colors, then the color rendering properties of the lamp are effectively comparable to another lamp with a very good R_a value.
3. Another aspect of importance is that a very high R_a value for a lamp only means that the colors perceived under its illumination are similar to the colors under the reference light source. Sometimes, however, colors that are slightly different from the reference colors are preferred. As the above examples show, the color rendering properties of such lamps cannot be judged correctly solely by the Ra value.

4.6.4 *Color appearance of light sources and perceived colors*

As described, the influence of the difference in the color temperature of the light sources on the perceived color is compensated, at least partly, by chromatic adaptation. Lamps of the same color temperature, however, can be made with many different spectral energy distributions. For this reason, the color rendering properties of lamps with the same color temperature can differ considerably because their spectral energy distribution is different.

4.6.5 *Color rendering and brightness*

The stimulus that yields the brightness sensation is the luminance. However, if one observes an object carefully, the perceived brightness of the object can be different, depending on its colors, even if they have the same luminance. Experimental observations revealed that the sensation of illuminated objects being lighter or darker in a room depends very much on the color rendering properties of the light source employed.[13] These comparisons were made in a room illuminated with fluorescent lamps of different R_a values and with an incandescent lamp with an R_a value of 100. The results are shown in Figure 9, in which the ratio of the illuminance under the incandescent lamp to that under the test fluorescent lamps to obtain the equivalent subjective brightness is plotted as a function of the R_a value of the test lamps. It is observed that, with decreasing R_a of the test lamps, the illuminance under those lamps necessary to obtain the equivalent subjective brightness increases remarkably.

4.7 Other chromatic phenomena

4.7.1 *Purkinje phenomenon*

This phenomenon was named after the Czech psychologist Purkinje, who discovered it in the early 19th century. The phenomenon concerns variation in the relative lightness of perceived colors between red and blue with changes in the luminance of the field of view. As the field becomes darker, the perceived red colors become relatively darker than the blue colors.

Chapter four: Color vision 133

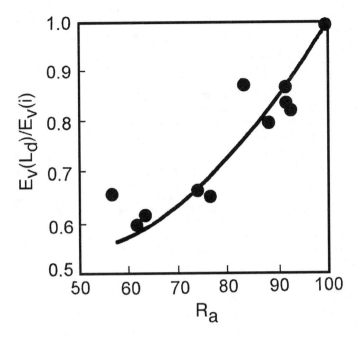

Figure 9 Relationship between the illuminance ratio to obtain equivalent subjective brightness ($E_v(L_d)/E_v(i)$) and the R_a of a test light source. $E_v(L_d)$: Vertical illuminance under an incandescent lamp with R_a of 100; $E_v(i)$: Vertical illuminance under a test light source. (From Kanaya, S., Hashimoto, K., and Kichizpe, E., *CIE Kyoto, 19th Session*, 1979, 274-278. With permission.)

4.7.2 Metamerism

Metamerism is a visual phenomenon in which two different colors (with different spectral reflection factors) are seen as the same color under a specific set of conditions. The condition requires a proper combination of spectral reflection factors of the object, of the spectral energy distribution of the lamps, and of the spectral sensitivity of the observer's eyes.

4.7.3 Bezold-Brucke effect

The Bezold-Brucke effect is a chromatic phenomenon in which the color of a light with constant spectral characteristics is perceived as a different color as the lamp's luminance is varied. A definite relationship between the perceived colors and the luminance has not yet been established.

4.7.4 Helmholz-Kouroshe effect

The Helmholz-Kouroshe effect is a visual effect in which a colored object is perceived to be lighter than an achromatic object with the same reflectance factor or the same luminance.

References

1. CIE, *The Basis of Physical Photometry*, CIE Publication No. 18.2, 1983.
2. Leonardo da Vinci, *Trattato della Pittura*, Langlois, Paris, 1651.
3. Newton, Sir Isaac, *Opticks*, Sam. Smith & Benj. Watford, London, 1704.
4. Young, Thomas, On the theory of light and colours, *Phil. Trans.*, 1802, 12-48.

5. von Helmholtz, H., Uber die Theorie der zusammengesetzten Farben, *Ann. Phys. Lpz.*, 87, 45-66, 1852.
6. Hering, E., Zur Lehre vom Lichtsinne Grundzug einer theorie des Farben sinnes, SB K, *Akad. Wiss Wien. Math. naturwise.*, K70, 169-204, 1875.
7. Muller, G.E., Zur Psychophysik der Gesichtsempfindungen, *Z. Psy. Physiol., Sinnesorg.*, 10, 1-82 and 321-413, 1896.
8. Walraven, P.L. and Bouman, M.A., Fluctuation theory of colour discrimination of normal trichromats., *Vision Res.*, 6, 567-586, 1966.
9. Ikeda, M., *Fundamentals of Colour Engineering*, Asakura Bookstore, 1980, 244-245.
10. CIE Publication in preparation, *Testing of Supplementary Systems of Photometry*, TC1-21.
11. Kelly, K.L., *J. Opt. Soc. Am.*, 33, 627, 1943.
12. Hard, A. and Sivik, L., NCS—Natural Color System: A Swedish standard for color notation, *Color Res. Appl.*, 6, 129-138, 1981.
13. Kanaya, S., Hashimoto, K., and Kichizpe, E., Subjective balance between general colour rendering index, colour temperature, and illuminance of interior lighting, *CIE Kyoto, 19th Session*, 1979, 274-278.

Index

Subject

A

Adsorption isotherm, 68–69

B

Bezold-Brucke effect, 133
Blaze angle, 5
Blaze wavelength, 6
Boxcar integrator, 32–33

C

Camera mirror, 4
CCD, *See* Charge-coupled devices
Charge-coupled devices (CCD), 12–14
  photodetector array, 32
Chromatic adaptation, 128–129
Chromaticity point of color, 121
Chromaticness index, 127
Chromatic phenomena
  Bezold-Brucke effect, 133
  Helmholz-Kouroshe effect, 133
  metamerism, 133
  Purkinje phenomenon, 132–133
CIE colorimetric system, 122–125
CIE UCS colorimetric system, 125–126
Color of light and color temperature, 128
  chromatic adaptation and colors of light sources, 128–129
  mired (micro-reciprocal degree), 129–130
Color rendering
  and brightness, 132
  color appearance of light sources and perceived colors, 132
  color rendering index
    general color rendering index (R_a), 132
    special color rendering indices, 132
  general color rendering index and perceived colors, 131–132
  method of measurement, 130–131
Colors and the color systems, specification of perceived colors, specification of
  Munsell color system, 130
  NCS, 130
  psychophysical colors, specification of, 121–122
    CIE colorimetric system, 122–125
    CIE UCS colorimetric system, 125–126
    ULCS colorimetric system, 126–128
Color vision, models of, 119–121
Color vision and eye
  retina, 116–117
  spectral sensitivity of, 117–118
Cones, 116
Correlated color temperature, 129
Czerny-Turner mount, 4

D

Dark current, 8
DC current-voltage converter, 15
Debye-Scherrer method, 72–74
Decay function, 31
Detector, phosphors, 31–32
Deuterium discharge lamps, 46–47
Discharge lamp, 18–19, 34, 45
  deuterium, 46–47
Doppler beat, 67

E

Electron-beam excitation, 22–25, 37–38
Excimer lamp, 22

F

Fluorescent lamps
  light flux of, 98–99
  light output of, 89–91
Fresnel's law, 107

G

Gravimetric method, 68
Green density, 75

H

Helmholz-Kouroshe effect, 133
High-energy beams, 23
Hydrogen discharge lamps, 18, 45

I

Image intensifier, 14–15, 32

J

Johnson's theory
   light flux of a fluorescent lamp, 98–99
   reflection, absorption, and transmission in single layer, 93–94
   reflection, absorption, transmission and emission in a single layer within multiple layers, 94–98
   white fluorescent lamp, application to, 99–102

K

Kozeney's constant, 70
Kubelka-Munk's theory, 79–80, 105
   basic equations and their general solutions, 80–82
   light reflection and transmission of powder layers, 82–84
   optical properties of phosphor layers
      cathode-ray excitation, 84–87
      light output of fluorescent lamps, 89–91
      ultraviolet excitation, 88–89
      X-ray excitation, 87–89
   scattering coefficient, measurement of, 91–92

L

Laser, 20–22, 34
Light and color
   perceived colors, 119
   psychophysical colors, 118
Light detector, 7–14
Light transmission method, 62, 64
Linear optical detectors, 32
Lock-in amplifier, 15
Logarithmic normal distribution, 55–56
Low-energy beam, 22–23

M

Mercury discharge lamp, 18–19
Metamerism, 133
Monochromator, 3–7
Monte Carlo method
   continuum model, 110–112
   features, 105
   particle layer model, 106–110
   for phosphor screens, 105–106
   problem related to, 111
Multistage cyclones, 66
Munsell color system, 130

N

Natural color systems (NCS), 130
NCS, *See* Natural color systems
Normal distribution, 55
Number-base distribution, 57

O

Opponent color theory, 119
Opponent signal, 121

P

Packing, definition of, 74–75
Particle packing, 74–75
   measurements of apparent density, 75–76
   measurements of fluidity
      motion angle, 76
      powder orifice, 76–77
      rest angle, 76
Particle size, methods for measuring
   image analysis, 57–60
      preparation of microsection, 57
   particle motion, analysis of
      centrifugal sedimentation method, 65
      inertia force method, 65–66
      laser doppler method, 66–67
      sedimentation method, 62–65
   scattering of electromagnetic waves caused by particles
      diffraction method, 72
      light scattering method, 70–72
      X-ray diffraction and X-ray scattering, 72–74
   surface area analysis
      adsorption method, 67–69
      transmission method, 69–70
   volume analysis of
      coulter counter, 60–62
      sieving, 60
Particle size and its measurement
   measuring particle size, classification and selection of method for, 57
   particle size and distribution
      function, 54–56
      mean diameter of, 54
      types of, 54
   shape and size of particles, 52–53
      circularity and sphericity, 53
      shape factor and effective and equivalent diameters, 53
Phosphors
   data processing and
      baseline correction, 40
      signal-to-noise ratio, improvement of, 40–41

spectral sensitivity correction, 38–39
measurements in vacuum-ultraviolet region,
    45–46
  excitation spectra, measurements of, 49–50
  light sources, 46–47
  monochromators, 47–48
  sample chambers, 48–49
reflection and absorption spectra of
  measurement apparatus, 27–30
  principles of measurement, 25–27
screens, 105–106
Phosphors, luminescence and excitation
    spectra of
excitation source, 16
  electron-beam excitation, 22–25
  ultraviolet and visible light sources, 16–22
luminescence measurements, suggestions
    on, 25
measurement apparatus
  light detector, 7–14
  monochromator, 3–7
  signal amplification and processing
    apparatus, 14–16
principle of measurement, 2–3
Phosphors, luminescence efficiency and
measurement apparatus
  electron-beam excitation, 37–38
  ultraviolet excitation, 35–37
principles of measurement, 34–35
Phosphors, transient characteristics of
    luminescence and
experimental apparatus
  detector, 31–32
  pulse excitation source, 34
  signal amplification and processing, 32–34
principles of measurement, 30–31
Photomultiplier tube, 7–9
Photon counter, 16
Photon-counting apparatus, 33
Photopic vision, 117
Pipette method, 64
Powder characteristics, 52
Pulse excitation source, 34
Purkinje phenomenon, 132–133

R

Reference colors, 122
Retina, 116–117
Rose's extinction coefficient, 64
Rosin-Ramler's equation, 56
Rowland circle, 6–7
R-R distribution constant, 56

S

Schott UG5 glass filter, 35
Schumann-Runge bands, 45
Schumann UV region, 45

Scotopic vision, 117–118
Sedimentation balance method, 62, 64–65
Sedimentation volume, 75
Seya-Namioka mounting, 48
Seya-Namioka mount monochromators, 47–48
Signal amplification and processing apparatus,
    14–16
Signal-to-noise ratio, 40–41
Single-channel and multichannel detectors,
    11–12
Small angle method, 74
Soft X-ray region, 45
Solid-state detectors, 9–10
Specific gravity balance method, 62, 64
Specific surface area diameter, 54
Spectral luminous efficiency function of eye, 117
Spectral order, 5
Spectral photon irradiance, 2
Spectral tristimulus values, 122–123
Spectrum irradiance, 2
Stokes diameter, 53
Stokes' shift, 27

T

Test colors, 122, 132
Transient recorders, 33–34
Tungsten lamp, 17, 38

U

ULCS colorimetric system, 126–128
Ultraviolet excitation, 35–37, 88–89

V

Vacuum-ultraviolet region, 45
Visible light, 8, 17–22, 118
Volumetric method (BET method), 68

W

White fluorescent lamp, 99–102

X

Xenon discharge lamp, 18

Y

Young-Helmholtz's trichromatic theory, 119